the **NO-NONSENSE** guide to

WATER

Maggie Black

D0558206

'Publishers have created lists of short books that discuss the questions that your average [electoral] candidate will only ever touch if armed with a slogan and a soundbite. Together [such books] hint at a resurgence of the grand educational tradition... Closest to the hot headline issues are *The No-Nonsense Guides*. These target those topics that a large army of voters care about, but that politicos evade. Arguments, figures and documents combine to prove that good journalism is far too important to be left to (most) journalists.'

Boyd Tonkin,
The Independent,
London

The No-Nonsense Guide to Water
Published in Canada by
New Internationalist Publications   and
401 Richmond Street West, Studio 393
Toronto, Ontario
M5V 3A8
www.newint.org

Between the Lines
720 Bathurst Street, Suite 404
Toronto, Ontario
M5S 2R4
www.btlbooks.com

First published in the UK by
New Internationalist™ Publications Ltd
55 Rectory Road
Oxford OX4 1BW
New Internationalist is a registered trade mark.

Cover image: Fred Hoogervost / Panos Pictures

Design by New Internationalist Publications Ltd
Typeset by Avocet Typeset, Chilton, Aylesbury, Bucks
Printed by TJ International Ltd, Padstow, Cornwall, UK

Library and Archives Canada Cataloguing in Publication

Black, Maggie, 1945-
The no-nonsense guide to water / Maggie Black.

(The no-nonsense guides)
Includes bibliographical references and index.
ISBN 1-896357-92-X

1. Water. I. Title. II. Series: No-nonsense guides (Toronto, Ont.)

GB661.2.B54 2004   553.7   C2004-904133-9

the **NO-NONSENSE** guide to

WATER

Maggie Black

About the author

Maggie Black is a writer and editor on international development issues, including water resources management and public health. She has written extensively on water and sanitation for UNICEF, is the author of reports for WaterAid, the Water and Sanitation Project of the World Bank/UNDP, the Global Water Partnership and the EC. She is also the author of *Water, Life Force*, a lavishly illustrated celebration of water, published by **New Internationalist Publications**. A book on the drinking water crisis in India's villages, *Water: A Matter of Life and Health,* will be published by OUP, Delhi in late 2004. Her visits to the Narmada Valley in India where large dams are being actively opposed, subsequently the subject of **New Internationalist** issue no. 336, 2001, helped to inspire an earlier title in this series, *The No-Nonsense Guide to International Development.*

Acknowledgements

The following people are acknowledged as having influenced the understanding of issues covered in these pages or having led the author along the watery writing path: Martin Beyer, Rupert Talbot, John Kalbermatten, Anil Agarwal, Alan Hall, Medha Patkar, Roland Schertenlieb, Uno Winblad.

Foreword

WATER AND CULTURE go together. Water shortage is not about mere failure of rain. It is about the failure of society to live and share its water endowment. This essential proposition, too often ignored, is where this *No-Nonsense Guide* begins.

We who live in the fast industrializing and imploding South must re-learn this principle. And fast. The management of water is today a make-or-break challenge. If we get our water practice and policy wrong we will see increasing destitution. At the same time, the opportunities are enormous. Water management is the starting point for eliminating poverty. Water security is the starting point for food security.

But to get the water-practice right we will first have to deal with the poverty of the professional mind – which has over time become fossilized and rigid in its outlook. We need a movement for water literacy, so that we can build a new understanding based on past traditions, current policies and future imperatives of managing and sharing a common water future. Contributions to building that movement – of which this book is one – are urgently needed.

Take the fascinating cases of ancient Rome and Edo (on which modern Tokyo is built). Romans built huge aqueducts to bring water to their settlements. These aqueducts are still the most omnipresent symbols of that society's water management. Many experts have praised the Romans for the meticulousness with which they planned their water supply systems.

But, no, these aqueducts represent not the intelligence but the utter environmental mismanagement of the great Romans. Rome was built on the river Tiber. The city did not need any aqueduct. But as the waste of Rome was discharged directly into the Tiber, the river was polluted and water had to be brought from long distances. Water outlets were few as a result and the élite appropriated these. By contrast, the inhabitants of Edo never discharged their waste into the rivers. Instead they composted the waste and used it in the fields. Thus, Edo had numerous water outlets and a much more egalitarian water supply.

We are also following Rome when we turn our backs on the water around us. Out of sight, of mind. Flush it and who cares. But care we must.

It is for this reason we need this *No-Nonsense Guide to Water*. We need to examine water as a fundamental life force, a maker and breaker of societies and dynasties, not only an aspect of public health or environmental integrity. This book charts a course through these shifting channels, offering its own take on current debates about 'water wars' and the corporate theft of the global freshwater commons. We learn desperately needed lessons that change mindsets, so that we can change policy and practice.

Water, the world's most fluid substance, disappears so rapidly. India is a crucible of the freshwater crises facing the world. Imagine you had a hectare of land in Barmer, one of India's driest places, and you received 100 millimeters of water in the year, common even for this desert region. The mathematics is simple: if you harvest just 100 mm of rainfall on one hectare, you receive as much as a million liters of water. If we don't capture the rainfall even the wettest place on earth will face shortages. Cherrapunji in the northeast of India has an annual rainfall of 14,000 mm, but it has chronic water shortages.

Water is about nothing less than our own and our planet's survival, as Maggie Black demonstrates in these pages. We have to take a no-nonsense view of this crisis and turn it into the opportunity of our lifetime. It is possible. And we must do it together.

Sunita Narain, New Delhi
Director of the Centre for Science and Environment.

the **NO-NONSENSE** guide to

WATER

CONTENTS

1 Water: life force

Water is key to the survival and growth of all human, animal and plant life and all economic and environmental processes. Not surprisingly therefore, water was worshipped in antiquity and plays a central role in myth and religion. Taming water to put it to humanity's service has been the stuff of heroic hydraulic undertaking for thousands of years. But the world's renewable and non-renewable supplies of freshwater are under increasing threat. There is even talk of 'water wars'. Is there really a world water crisis? And what is it actually about?

'WATER WAS THE matrix of the world, and of all its creatures.' Most books on water start with a quote. Or with folksy accounts of bubbling springs or pumps in the back yard – but let's leave those to one side.

This quote is from a 16th century German physician, Theophrastus Bombast von Hohenheim known as Paracelsus,[1] and claims as much for water as can be claimed for anything. His view of water's fundamental significance to life on earth was neither unique nor original. In many mythical versions of the world's beginnings, water only just plays second fiddle to God. 'Darkness was there, all wrapped around by darkness, and all was Water indiscriminate' runs the creation hymn of the Rig Veda, written around 3700 BC. In the Old Testament book of Genesis, earth was a void without form until 'the Spirit of God moved upon the face of the waters'. Variations on the theme of water as the source of life are present in virtually every religion.

Today, those of us who live in water-rich lands take too much for granted this fluid as physiologically vital as blood, at once the giver of life and its frequently temperamental destroyer. Our forbears with their sacred springs, holy wells and rain gods did not make such an error. They understood water's life-giving role and honored it in their celebrations and rites. Its other mysteries took time to unravel. This most ordinary of fluids – flavorless, odorless and colorless – has been a source of scientific fascination for centuries, and its mysteries still puzzle analysts today. Leonardo da Vinci was the first to investigate water's physical properties. His book, *Del moto e misura dell' acqua* (Of the Motion and Measurements of Water) described the experiments he undertook to work out how and why running water behaved the way it did.

This fusion of two parts hydrogen to one of oxygen – we all know the formula although how atoms of gas combine to make a liquid is beyond most of our comprehensions – cannot be fixed in any one place. Nor does it assume a permanent shape. Water feels free to ignore all chemical rules. At different temperatures and under different pressures, it changes form and weight and becomes different things: mist, snow, ice, steam, cloud, drops, spray, spume. It also blends with other substances, transporting them away, absorbing or dissolving into them, even corroding them, making something new. And it has extraordinary natural self-cleansing powers. All of these characteristics contribute to water's unique life-generating force.

We know, because we learned it in primary school, that without water nothing on earth could survive or grow. But we rarely reflect that water powers our bodies and minds more elementally even than food, of which it is a key constituent. We are made of water, more or less: 70 per cent of our tissues and 55 per cent of our blood. Two-thirds of the planet's surface is covered by water. On land, freshwater sustains the ecological balance necessary for planetary health, flowing through the network of veins and arteries in the earth, lubricating its soils, shaping its cavities and depressions, and fuelling its store of fertility. The loss of our water supply – a crash in the hydrological cycle – would wipe us out more thoroughly than the explosion of any nuclear arsenal.

Thus to control water is to control life. No wonder that, down the ages, kings and presidents went in for taming rivers and damming lakes. Hydraulic or super-natural power over water – they were sometimes con-fused – commanded respect for several millennia, making and breaking dynasties from the Euphrates to the Nile, the Ganges to the Indus, the Yangtze to the Po. But the nature of that power and the foundations on which it rests changed down the years. Science and technology took over where primeval forces once ruled. The exercise of power in the modern world is a cerebral affair, a manipulation of economic and polit-ical forces which has little in common with the ancient battle against the elements.

Today, pressures are at work to confront us once again with the reality of water as the 'matrix of the world'. According to the hydrologists, we are facing a deepening freshwater crisis. The gauge on the world's water tank is dipping dangerously low. For too long, we treated water as an abundant and limitless free resource and squandered it recklessly. Now, as both populations and their per capita consumption of water remorselessly rise, we are beginning to rue the consequences. Ominous conflicts over water are on the way – in some places, are already occurring.

Important questions need to be answered. Who owns the bubbling spring and who gets to drink from it? Are the many millions of pumps today extracting water from beneath the ground doomed to run dry? Believe me, this is not a folksy story.

The global water-pot
Before examining the scarier scenarios, it is worth inspecting the contents of the global water-pot. This is a source of magnificent statistics – about volumes measured in cubic kilometers (km^3), thundering flows, rates of evaporation, ice-caps, vast river net-works and underground rock pools too deep to mine. Inventories of global water supplies vary by amounts equivalent to the size of several small seas.

The best-educated guess about the total global

potful – from the State Hydrological Institute in St. Petersburg which undertook a recent inventory of the world's water for UNESCO[2] – is that there are 1.4 billion km^3 of water on earth, liquid and frozen. Hardly any of this is useful for human consumption: 97.5 per cent is in the oceans and is too salty to drink or use for irrigation. Of the remaining 2.5 per cent, two-thirds (or 24 million km^3) is locked up in the polar ice-caps and permanent snow cover. This amount changes over time. About 18,000 years ago, ice covered two-thirds of the earth's surface. That has now reduced to 12 per cent, and continues to shrink – rapidly under the influence of global warming.

A small amount of the global freshwater supply is in the air at any time, in the form of rain, clouds, or vapor. Another small amount is contained in living things – plants, animals, human beings. A large proportion of the remaining 16 million km^3 lies imprisoned in nooks of sedimentary rock too far underground to exploit. Freshwater lakes and rivers, from which we obtain most of our water, contain only around 90,000 km^3 – only 0.26 per cent of the global supply. If the amount they contain was spread evenly over the globe, it would make a layer 1.82 meters deep.

This water is far from static – indeed, if it stood still it would de-oxygenate and become foul or unusable. Instead, the water in streams and on the surface of lakes flows constantly downhill, eventually reaching the sea. By the phenomenon known as the hydrological cycle, it is constantly renewed and put back where it began. Every year, a vast quantity of water – over 500,000 km^3 – evaporates from the seas courtesy of the sun. Sucked up as high as 15 km from the world's surface and transformed into clouds, two-thirds of this water is returned to the oceans as rain or snow.

The rest falls on land, conveniently scrubbed clean of the impurities it picked up the last time it ran about on earth. Back on the ground, it finds its way into the shallower underground aquifers, or enters streams

The hydrological cycle

In a year, the sun's heat evaporates the equivalent of a layer of water 125 cm deep off all the world's oceans. Two-thirds falls back into the sea. The rest drops from clouds onto the land, penetrating the soil and making its way into aquifers, streams and lakes. Plant growth releases water into the air, and rivers flow into the sea, thus completing the cycle. Over millennia, vast quantities of freshwater have built up in the polar ice-caps.

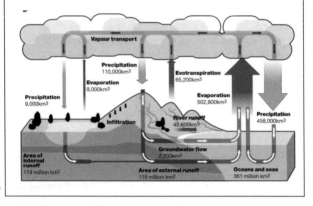

H₂O, Guardian Special Supplement, 23 August 2003

and rivers where it again embarks upon a journey to the sea. The process is simple enough in its basics to explain to a child, but its elements are extremely complex. Cogs in the hydrological cycle turn at many different speeds. Evaporation from the sea takes only 10 days to re-appear as rain. If it lands in the upper reaches of a river basin it will take some weeks to reach the sea. By contrast, water locked up in polar ice and glaciers has been frozen for hundreds of thousands of years. And the deepest groundwater aquifers now being tapped contain water that has taken millennia to accumulate.

Some of the water now being consumed from the global pot is therefore 'non-renewable'. Once mined, it has effectively gone for good. But the rest – the 'run-off' landing on the earth – provides the world with its renewable supplies. Capturing this run-off has been the task addressed by the hydraulic engineers of antiquity and all of their successors. A Sinhalese King, Parakrama the Great (1153-1186 AD), a monumental builder of artificial lakes in Sri Lanka, declared: 'Not a single drop of water received from rain should be allowed to escape into the sea without being utilized for human benefit.'[3] This is the sentiment which has inspired dam- and reservoir-builders down the years, with – it has to be admitted – extraordinary benefits for humankind, whatever today's dam-related misgivings. King Mahasena, a third century lake-building precursor of Parakrama, was seen as so beneficent a savior from hunger and thirst that he was deified by his people as a god.

The amount of run-off theoretically accessible via surface and underground streams is estimated at around 34,000 cubic kilometers a year. But much of this occurs far from human settlements. The World Meteorological Organization therefore states that a more realistic figure of readily available freshwater is around 12,500 cubic kilometers.[4] This ought to be enough to go round: we are still only using around half of it. Most regions and countries have well above the notional volume of 1,700 cubic meters per head per year regarded as the cut-off between water comfort and water stress. No-one, even in water-profligate USA and Canada where volumes are used on swimming-pools, golf-courses and in all sorts of inessential ways, remotely gets through that quantity a year.

But all of these figures are an illusion if you look at the realities on the ground, in the same way – only more so – that Gross Domestic Product (GDP) divided by population gives no sensible grasp of the average person's income. The renewable supply of water is not distributed evenly, either from hill to valley or around the globe. Far from it. The uneven disposal of water about the earth, rather than the contents of the total pot, coupled with rapidly increasing rates of extraction from both renewable and non-renewable sources, is an important cause of today's freshwater trouble.

Unjust rainfall

Shakespeare wanted us to believe that rainfall was an egalitarian affair, landing equally upon the just and the unjust. Perhaps so if they are standing close together. But at larger distances, there is nothing fair about who gets rained on and who does not. You could spend a hundred years in the middle of the Sahara desert and collect less rain than on a single day in Hawaii. And depending which side of the mountain you are on – the side where the clouds deposit their load or on the other side in the 'rain shadow' – the contrast can be striking. On the big island of Hawaii, annual rainfall varies from 5,100 millimeters on the side of the island facing the Trade Winds, to less than 250 mm on the downwind coast.[5]

The global annual rainfall average is 760 mm. This amount is typical in western Europe and over the US and Canadian prairies. Other regions – for example Australia, northern China and the Middle East – receive much less, and some tropical environments receive much more. Deserts typically receive less than 100 mm; much of the Sahara, the Namib in South-West Africa and the Peruvian seaboard receive far less even than this. Some rain-short areas are among the coldest on earth – Arctic North America, Greenland, and Mongolia. Life in dryland places – from Los Angeles to Timbuktu, from Damascus to Australia's Murray Darling Basin – is entirely dependent on the careful management of scant supplies.

Many parts of Asia receive luxurious quantities of rain: most of Indo-China, Indonesia and Philippines, for example. In other parts, there are wide rainfall variations. Cherrapunji in North-Eastern India near the Bangladesh border is famously the wettest place on earth with 13,390 mm – 13.4 meters – of rain a

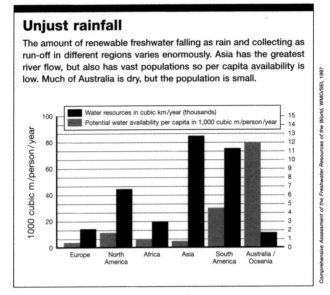

Unjust rainfall

The amount of renewable freshwater falling as rain and collecting as run-off in different regions varies enormously. Asia has the greatest river flow, but also has vast populations so per capita availability is low. Much of Australia is dry, but the population is small.

■ Water resources in cubic km/year (thousands)
■ Potential water availability per capita in 1,000 cubic m/person/year

Europe · North America · Africa · Asia · South America · Australia / Oceania

1000 cubic m/person/year

Comprehensive Assessment of the Freshwater Resources of the World, WMO/SEI, 1997

year. Meanwhile, over in the Thar desert on the India-Pakistan border, years pass without a single drop.

In this part of the world the most important rainfall phenomenon is the monsoon. Unlike in temperate zones, where some rain descends at all times of year, in a monsoon climate all the year's supply arrives during one or at most two rainy seasons. In drier areas, the total year's supply may land within a few hours. Farmers rush to plant their crops before the rains disappear. Hence the desperation to trap temporarily swollen streams behind bunds and dams, slow them down and let run-off seep into the soil, saving the rain for later use.

In wetter, more tropical areas, downpours may be so tempestuous and of such a volume that rivers burst their banks, crops are washed away, and floods destroy the very life the water was supposed to nurture. China, the most water-fraught country in the world, has suffered appalling floods throughout its history: the swollen waters of the Yellow and Yangtze Rivers have drowned millions down the years, often precipitating famine. The catalogue of flood devastations continues today, with the fear that climate change and the environmental degradation caused by tree-cropping is exacerbating them.

At the opposite extreme is drought, coupled inexorably with famine. The images etched on our consciences are of parched earth, withered crops, livestock carcasses, and hollowed human forms plodding wearily along scorched African paths in search of food relief. But the significant question is less how low is the rainfall than how it compares to the local average. In Mongolia, 500 mm would be seen as manna, but in Miami or Manchester it would be a disaster. In temperate Europe where agriculture is rain-fed, a period of several weeks without any rainfall constitutes a potentially serious drought. 'Pray for rain' might as well be scrawled on the wall of a Moroccan village as on a truck in Texas or Afghanistan.

Flora and fauna, as well as human settlement and livelihood patterns, adapt to whatever is the norm. In Lima, capital of Peru on the Pacific seaboard, people start to build upper storeys on their houses when they can afford to and no-one worries about the roof: it never rains. In Bangladesh, lashed by storms in the Bay of Bengal and watered by huge quantities of rain and snowmelt descending from the Himalayas via the Ganges and Brahmaputra, catastrophe is only announced when one-half of the country is flooded. But if less than one-third is under water, that is a disaster of another kind. The most densely occupied rural landscape in the world depends on the annual inundation for its rice-bowl fertility.

On the vagaries of rain depend the methods of agriculture and the type of crops people plant. Where rainfall is abundant, it is used to cultivate crops such as rice, for which 3,450 liters is required to produce just one kilo.[6] The staples traditionally cultivated in arid areas are less thirsty varieties of grain: sorghum and millet, for example. In western and central Africa, rural people still depend heavily on cassava, known as a 'famine crop'. This starchy root may not be highly nutritious, but it requires little rain, grows in poor soil and can lie in the ground until it is needed.

In most deserts, livestock herding is the only practicable livelihood. At least the 'crop' can walk to wherever rain has replenished the grazing and re-filled the water-holes. In central Australia, lands stocked are so marginal that their carrying capacity is no more than one head of cattle per square mile (1.6 sq km).

Mighty rivers

Rivers are arteries vital to the life of the lands through which they flow. Some of the greatest rivers extend thousands of miles, linking environments of great diversity: at the source, lush vegetation; at the mouth, vistas of sand. Rivers are important highways through the landscape, seeking out softer rock formations to carve their way through mountains and meander through gentler terrains.

Wherever they go, rivers provide a natural pipeline for the rainwater deposited on earth. Some do this more equitably and with more generosity to humanity than do others. The Amazon, flowing in one of the wettest parts of the world, carries 16 per cent of global run-off. The Congo basin carries one-third of the river flow in all of Africa. Along with the water come particles of soil and rock, organic matter and dissolved plant nutrients, washed into the flow and transported to its quieter lower reaches. And of course wastes.

All rivers are part of a network in which smaller tributaries drain into the larger stream. Small river valleys, in which human settlements congregate, connect within a larger river basin or catchment. The successive merging of watercourses makes up a system which, in the case of rivers such as the Mississippi, the Danube, the Ganges and the Rhine, cover hundreds of thousands of square kilometers. By the time great rivers reach the sea, their mouths are often several kilometers across. Some delta areas are vast networks of islands, streams and low-lying banks which change their topography every year. Holland and Bangladesh are classic examples of countries sitting on delta lands, whose riverine nature is the defining feature of their environment and the society it has fostered.

Some rivers – for example, the Nile, the Niger, the Jordan and the Colorado – receive much of their flow from lakes or springs at very distant sources. During their voyages they encounter large quantities of rainless desert lands and suffer huge losses from evaporation. The Niger only avoids running dry because it turns southwards into humid savannah country. The Nile survives its journey across the Nubian and Arabian deserts only because the abundant Blue Nile joins at Khartoum. For the last 2,700 km of its course it flows through completely rainless desert and without any tributary to swell it.[7] Without the Nile, Egypt – which is virtually rain-less – could not exist.

Many less fortunate streams give up. Rivers that only flow seasonally are 'ephemeral': after a burst of rain at their upper reaches, they fan out and dry up, leaving rock-strewn pathways, ending in swamps or salt-pans. This type of stream is common in Australia and Africa. Namibia, with the driest climate south of the Sahara, is almost entirely dependent on ephemeral rivers.

Where rainfall is limited or seasonal, water in surface streams or underground aquifers offers the only means of cultivating crops. Irrigation systems for manipulating water about the landscape using draught animals, buckets and wheels have been in operation since time immemorial. The agricultural output of many countries and large areas within them – India, China, Egypt, California, Bangladesh, Pakistan – depends on irrigation from rivers and renewable underground sources.

The world's rivers great and small are therefore immensely diverse in their make-up, their flows, their geographical spread, their water levels, the aquatic and human environment they support, and their seasonal behavior. They are also anarchic: they disrespect political and administrative boundaries. Worldwide, 269 river systems cross national borders, and the contest between upstream and downstream users – and polluters – can be bitter, especially as pressures on their waters increase. According to the UN, 158 international river basins could prove to be flashpoints for future conflict. Tensions between Israel and its neighbors, Turkey with Syria and Iraq, Egypt with Sudan and Ethiopia, Bangladesh with India and Nepal, continually arise over their dependence on rivers swollen far away with someone else's rain.

Rivers are not only a vital source of water for agriculture and all domestic purposes. Many other livelihood systems, especially fishing, are based on rivers. When their flows or volumes are altered, or they are corralled for the exclusive use of upstream users, the impact on the riverine environment and those who depend on it for a living can be dramatic. One of the most extreme examples is the case of the Aral Sea. Since 1960, heavy withdrawals of water from its feeder rivers to provide irrigation have caused what used to be the world's fourth largest water body to shrink to a quarter of its size.

The interaction between rivers and the landscape has affected every single aspect of human life and livelihood since civilization began. Alterations in those dynamics have caused the rise and fall of dynasties and economic regimes. The pace of those alterations dramatically speeded up in the 20th century. What was originally seen as a triumph of technical ingenuity over nature now causes increasing alarm. But not enough to halt those addicted to the grand slam version of 'development' progress.

Aral: A sea ruined beyond repair

What hydraulic lunacy has done to the Aral Sea is the ultimate ecological cautionary tale. Once the fourth largest inland sea on earth,
since 1960 the Aral has shrunk to just a quarter of its original size.
The sea is the depository for two major rivers, the Amu and Syr,
descending from the Himalayas into Central Asia. Soviet planners
thought their waters were running to waste. In the 1950s, they built
canals near the border with Afghanistan and diverted vast flows to
grow cotton in near-desert terrain.

The Aral soon began to dry up. By 1970, the village of Muynak,
once an island in the Amu delta, was 10 kilometers from the coast;
by 1980, the distance had grown to 40 km; by 1998 to 75 km. As the
waters shrank, the greater concentration of salt killed all the fish; by
1977, the catch – 50,000 tons a year – was reduced by 75 per cent.
Within five more years the entire fishing industry had closed down.

There were other effects. The water table in the area sank, destroying many oases. Unirrigated farms dried up. Wildlife disappeared,
wrecking the fur industry. Over-irrigation leached ever more salt to
the surface, and the build-up of salts and pesticides on 28,000 sq km
of exposed sea bed became a stew of poisonous sediments. Dust
storms blow these over farmlands and into people's lungs, causing
cancers and respiratory ailments.

The climate has also changed. The sea acted as an air conditioner
in summer and a radiator in winter. The climate is now continental,
with wider extremes of temperature, less rainfall, and a shorter growing season – making it impossible to grow cotton because it needs
time to mature. So farmers switched to even thirstier rice.

By the late 1980s, the Soviet authorities realized that an ecological
catastrophe was in the making and instituted a disaster mitigation
plan. But little had actually been achieved before the USSR collapsed, leaving the crisis of the Aral in the hands of five brand new,
cash-strapped and mutually antagonistic Central Asian nations. In
1992, they set up a basin organization and pledged to co-operate to
restore the sea. Although they have agreed a joint plan for water conservation and related measures, skepticism remains that its provisions will ever be assiduously applied.

The Aral Sea continues to shrink, if a little less quickly than it did
before, and the pollution remains as severe, coating the land with
white toxic dust for miles in every direction. Ecological tourists come
to stare in horrified fascination at ships disintegrating in the landscape and the bleached bones of poisoned livestock protruding
nearby. Some believe there is less than 10 years left before the whole
area becomes a desert. It is already the most complete example on
earth of how bad hydrological science can combine with political
inertia to produce large-scale environmental disaster.

Civilization follows water

The evidence of water's significance in the way the world developed is all around us. Every city and large town in Europe grew up around a river: Rome on the Tiber, Paris on the Seine, London on the Thames. The very earliest civilizations were established in river valleys: Mesopotamia around the Tigris and the Euphrates, Egypt on the Nile, the dynasties of Xia and Shang along the Yellow River. Often, major settlements were at the coast or in the rivers' lowest reaches, so that they could enter the sea and send outward fleets of ships for purposes of trade or conquest.

All these civilizations had to be able to manipulate water. The Harappan civilization in the Indus Valley of 3000-1500 BC was one of the very earliest to build towns and trade internationally. Without sophisticated water management techniques, this would have been impossible. Wells and irrigation works on the Indus can be dated back to 2,600 BC and archaeologists have found evidence of the use of lift and surface canals for irrigating winter crops.[8] Even in such distant times, people realized that if water descended from the skies only in violent bursts during a part of the year, ways had to be found to capture and control it so that its life-giving benefits could be prolonged.

So prolific are the examples of well-drilling, underground aqueducts, and waterworks constructed on a monumental scale, from Mesopotamia, Persia, Egypt, the Indian subcontinent and China, that one historian – Karl Wittfogel – suggested that the empires of the ancient world depended on hydraulic might.[9] Their centralized power was built on control of great rivers, harnessed by feats of engineering. Dams, tanks, reservoirs – even hanging gardens – were built by huge slave labor forces and managed by bureaucrats. These water works, on which depended the livelihoods of the empire's inhabitants, created the opportunity for despotic patterns of government and society.

Where water is scarce and naturally precious, there

is undoubtedly a strong connection between power over water and power over people. Down the centuries, kings, *maharajahs*, even Prime Ministers and Presidents, have identified themselves with vast hydraulic projects. When the 221 meter Hoover Dam on the Colorado River was inaugurated in 1936, it was the highest in the world and truly awe-inspiring. By then Herbert Hoover's engineering supremacy had taken him up the ladder of political power, into the White House, and – as it happens – out again, in the wake of the Great Depression.

The leitmotif of water as power was later adopted by several leaders of newly-independent nations. As he walked round the site of the massive Bhakra-Nangal dam in 1954, Prime Minister Jawaharlal Nehru remarked: 'These days, the biggest temple and mosque and *gurdwara* [Sikh temple] is the place where man works for the good of mankind. Which place can be greater than this?' Within a few years President Abdel Nasser of Egypt had begun to build the Aswan High Dam. President Kwame Nkrumah of Ghana's grandiose ambitions included a high dam on the Volta and the creation of the largest reservoir in the world. Mao Ze-Dong dreamed of re-shaping the Yangtze River, and the result decades later is the Three Gorges Dam.

The Chinese are not the only water despots in a long inherited tradition. In 1993, Saddam Hussein set his engineers to drain Iraq's southern marshlands, displacing 500,000 Marsh Arabs and destroying their way of life. This was a throwback to biblical leaders such as King Nebuchadnezzar of Babylon, for whom hydraulic construction was the ultimate expression of might. Thus civilizations are not only fashioned by water, they can also be destroyed by its control in centralized and undemocratic hands.

Natural forces may initially decide where water flows, and who is graced and who deprived by the arbitrary behavior of weather, climate and hydrogeology.

Beyond that, political and economic considerations prevail. Ultimately, who receives water and who does not is a question of how humanity's technological prowess is deployed to manage and distribute the available freshwater supply.

So is there a world water crisis?

There is not one world water crisis, but several. The idea of a world running dry is not convincing, and the notion of 'water wars' engulfing the 21st century has been much derided. But the water stress already being experienced in many countries and neighborhoods is real enough. In most cases, the paucity of their natural freshwater endowment is an important part of the picture. Growing populations and food demands, increased pollution, and a more water-profligate lifestyle, fills that picture out.

The water quantity problem countries are mostly in Africa and Asia. One-third of Africa's population – 300 million people – already live under conditions of scarcity, and this number is likely to increase to more than a billion by 2025. Nine of the 14 countries of the Middle East are in the same boat, and six of them will double their populations within 25 years. Both India and China are also in trouble. India, with 20 per cent of the world's population, has to make do with 4 per cent of its freshwater resource; water riots and disputes over river take-offs are becoming more strident every dry season. China with 22 per cent of the world's people already barely manages with 6 per cent of its water: 300 towns suffer regular shortages.

Growing pressures on the environment have led to another kind of water crisis: pollution. The self-cleansing powers of water are being overstretched by the volume of wastes – human and chemical – deposited in them directly, or carried into them by rainfall run-off. Many of the rivers of Eastern Europe are filthy. Three-quarters of Poland's rivers are so contaminated by chemicals, sewage and agricultural run-off that their

More people, less water

Growing populations are an important factor in water stress. 500 million people were living in countries chronically short of water in 2000. By 2050, the number is expected to rise to 4 billion, around 40 per cent of the world population. Consumption of water – in industry, agriculture and domestic use – is also rising. In 1900, around 350 cubic meters were used per head; by 2000, this had risen to 642 m³.

Have and have nots

Percentage of the world's population with different water availability *2000*

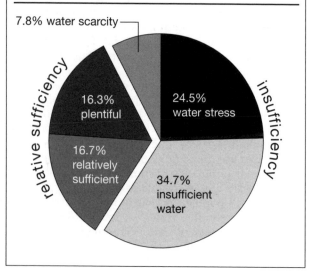

7.8% water scarcity

16.3% plentiful

16.7% relatively sufficient

24.5% water stress

34.7% insufficient water

relative sufficiency

insufficiency

Robin Clarke, Jannet King, The Atlas of Water, Earthscan, 2004.

water is unfit even for industrial use.[10] The shores of the non-tidal Mediterranean are awash with detritus. But the situation is worst in less-industrialized countries, where 90 per cent of waste water is untreated. In Latin America, 98 per cent of domestic sewage is discharged raw into the nearest stream; all of India's major rivers are similarly polluted. Human excreta is extremely pathogenic; its bacteria and viruses are responsible for outbreaks of deadly diarrhoeal infection, including cholera.

The groundwater picture is, if anything, more depressing. Always trumpeted as naturally clean and pure, groundwater is now the source of drinking water for one-third of the world's people. But aquifers close to the sea are being over-pumped, and the salt-water intrudes and makes it unusable. Chemical pollution of underground water is common in industrialized countries as well. Around many major cities, near industrial sites and under the soils of farms using quantities of chemical fertilizers and pesticides, groundwater has become seriously degraded. Unlike rivers, groundwater once poisoned cannot be cleaned.

The proposed solution to this hydra-headed crisis has underlined the reality that the control of water is ultimately about power. Ever since the early 1990s, when the pressures on the world's finite supply of freshwater first came into focus, the blame has been laid at the door of humanity's wasteful and extravagant misuse of an undervalued fluid. It follows that the remedy is to slap a far higher price on water – a price which reflects the costs of capturing it, channelling it, pumping it, dispersing it across cropland, using it as a transport medium for waste, de-polluting it and the rest. The mechanisms of supply and demand will mop up the problem, leading to changes in domestic, agricultural and economic behavior whereby water will be used more carefully – not squandered against the threat of a waterless tomorrow.

A watery crusade
This solution, not surprisingly, has met with considerable resistance. Not least because it plays straight into the hands of commercial enterprises looking to water for their next big opportunity. The entry of international market-place economics into the water arena has led just where one might expect. Not just the

pipelines, pumps, drains and treatment plants are now – with good reason in many cases – to be paid more for. The rain itself, where it lands on the earth, sinks down into an aquifer or flows into a stream or a river, is being co-opted for profit.

A resource traditionally regarded as a public good over which no individual may claim exclusive proprietorship is being monetized and treated as a corporate asset. Water is in this sense the latest or even the last frontier: among natural resources only air still escapes the relentless reach of capitalization and 'efficient utilization' by handy introduction of the profit motive. As in every case where natural resources are commandeered in the name of being more efficiently managed, those whose livelihoods depend on their use in traditional types of economy lose out heavily in the process.

Under the pressures of today's water crisis, the politics of water, from the local to the international level, are becoming extremely tense. It is becoming increasingly difficult to carry out massive schemes of water manipulation with all their accompanying re-arrangement of the landscape and its people: they provoke serious popular unrest. Protests are proliferating, against the privatization of services, against overweening hydraulic enterprises, against the pollution which is putting countless lives and habitats at risk. All these aspects – environmental, political, economic, social, ideological – need to be taken into account in exploring the parameters of a crisis placing life itself at risk.

That exploration is what this book attempts. And since water to drink is at the top of the list for humanity's survival, that is where the voyage will begin.

1 Philip Ball, *H₂O, A Biography of Water,* Weidenfeld & Nicolson, London, 1999. **2** Marq de Villiers, *Water Wars: Is the world running out of water?* Weidenfeld & Nicolson, London, 1999. **3** J.B. Disanayaka, *Water Heritage of Sri Lanka*, University of Colombo, 2000. **4** *Comprehensive Assessment of the Freshwater Resources of the World*, Swedish Environmental Institute, 1997. **5** Robert Henson, *The Rough Guide to Weather*, Penguin Books, London, 2002. **6** Peter Gleick, *The World's Water 2000-2001*, Island Press, Washington, 2000. **7** Eberhard Czaya, *Rivers of the World*, Cambridge University Press, 1981. **8** Anil Agarwal and Sunita Narain (eds), *Dying Wisdom: Rise, fall and potential of India's traditional water harvesting systems*, State of India's Environment 4, CSE, New Delhi, 1997. **9** Karl Wittfogel, *Oriental Despotism*, 1957; quoted in Colin Ward, *Reflected in Water: A crisis of social responsibility,* Cassell, London, 1997. **10** Maude Barlow and Tony Clarke, *Blue Gold: The battle against corporate theft of the world's water,* Earthscan, London, 2002.

2 Water, survival and health

The statistics say that globally 1.1 billion people are without water and 2.4 billion without sanitation. What does this mean? And how does it translate into risks of disease? The World Health Organization says that 80 per cent of illness is water-related, but poor sanitation and lack of hygiene are really to blame. The drive for public health prompted decades of effort to provide subsidized water supplies and sanitation. But many systems have failed and the impact on disease has not been impressive.

WHAT WORSE PROBLEM could anyone face – and 1.1 billion people supposedly do so – than lack of water?[1] Not to be able to drink, cook, bathe, wash clothes, flush waste away …

In fact, of course, no-one is without a water supply or they could not survive. In rural areas, where most of these 1.1 billion people live, the problem is that their water comes from a natural source such as an open well, pond or stream; in town, those without household connections rely on a street tap, pump or bucketfuls bought from a vendor. The task of household water collection invariably falls to women. They may have to walk long distances to reach the water source, and carry it home in heavy containers on their heads or backs, picking up potential contamination on the way. Obviously, all households want a dependable water supply within easy reach, and if nature does not provide it, then a tap or pump nearby is essential. To be safe to drink, the supply needs to be captured in a pristine condition, and come into contact with nothing impure on its journey into household vessel.

Water is more fundamental to survival than having somewhere hygienic to expel bodily wastes. So the 1.1 billion without water supplies tend to attract more urgent attention – and funds – than the 2.4 billion without sanitation. Both these statistics are unreliable.

Drinking water: what type of source?

People traditionally took their water from streams and open wells. These are unsafe or 'unimproved'. Improved sources include household taps, standpipes, handpumps, encased wells and springs, rainwater collection systems; all require construction and maintenance. They bring water close to the house; where they bring it inside they increase consumption dramatically, from a few to scores of liters a head per day, and much more washing and bathing gets done.

Percentage of population by type of access to water 2000

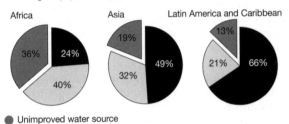

● Unimproved water source

Improved water source: ● household connection ○ other access

Excreta disposal: adequate facilities

Disease is usually to do with poor hygiene and failure to confine wastes containing dangerous bacteria. Piped water cannot be used for flushing in low-income areas: sewerage is far too expensive. So toilets in these settings are 'dry': wastes are confined to closed pits until they are 'safe' (and may then be recycled as fertilizer). 'Adequate facilities' therefore keep raw excreta out of reach. Pour-flush pans with a water seal are one kind of 'improved' toilet; the VIP – ventilated improved pit – is used in arid areas. Sanitation, even of the simplest kind, is invariably more costly than water provision.

Percentage of population by type of sanitation 2000

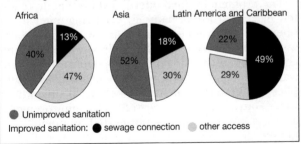

● Unimproved sanitation

Improved sanitation: ● sewage connection ○ other access

UNICEF, State of the World's Children 2004 and Robin Clarke, Jannet King, The Atlas of Water, Earthscan, 2004

Few non-industrialized countries really know how many of their people have functioning taps and toilets, and definitions of what constitutes service 'coverage' vary so widely as to render the data spurious. However, since it gives a ball-park view of the problem, we will use it.

Around 2 billion rural dwellers lack toilets, many more than urban, but their absence is more of a health hazard in town than in the countryside. 'Sanitation' can mean street drainage and rubbish collection too, but it always includes toilet facilities of some kind. Not necessarily a flush toilet – a water-wasting luxury device – but some place of excretion more hygienic than the proverbial bucket or a squat in the fields.

In the traditional world, freshwater was regarded as a free, communal resource and its capacities to clean, purify and refresh were revered. Many religious practices have a public health dimension. In Christian baptism, water washes away sin, and Muslims have to perform ablutions before they say their prayers. In some societies, a glass of water is still offered to the guest as the first mark of hospitality. In the deserts of Somalia and the Middle East, no-one is denied access to a water-hole to drink and water their cattle. The idea of water being a life-enhancing blessing provided by God for everyone to share has a long history. And for much of that history, water supplies were free, relatively abundant and normally safe to drink, even if collecting them was burdensome and inconvenient.

The problem in the modern world is that there are few places left where naturally occurring water in an open well or stream is regarded as safe to drink. 'Safe' water is engineered, and the engineering part of the equation is costly. The millions of people without safe water are actually people without *engineered* water supplies – or in development parlance, are people whose water sources are 'unimproved'. The 'improvement' gets costlier and more sophisticated as numbers of people grow, consumption of water rises, pollution increases and reliable sources are deeper down or farther away. The ways of the past, when drinking sources were natural, 'safe' and 'free' are vanishing.

Enter the public health engineers

As towns and cities grew in size during the modern era, they became unwholesome and insanitary places. Epidemics of plague, and later cholera, made regular visitations to London, Madrid, Paris and Rome. The countryside with its fresh air was where people went to be healthy. Rapid urbanization in 19th century Europe and North America accelerated the effects of crowdedness and urban squalor until they became notorious. In Britain, the first legislation to revolutionize town drainage and water supplies was introduced in 1848. At the height of that year's cholera epidemic, 3,000 Londoners died in a week. Parliament was finally goaded to take action when the appalling stench of the Thames outside their windows made it impossible for Members to sit.

The 19th century sanitary revolution brought something new to the water and sanitation story: the modern theory and practice of 'public health'. A turning-point was the moment in 1854 when the handle of a pump in the Soho district of London was locked on the advice of Dr John Snow. At the time, the prevailing wisdom was that cholera was contracted from the 'miasma' of the air. But Snow traced all the cholera cases in Soho to their source of origin: suspect water in a common pump. His discovery connected public health irrevocably to drinking water – not to the far guiltier excreta whose seepage into the supply had created the fatal chain of infection.

There began a monumental building program of pumping stations, water pipelines, sewers, drains and water treatment plants by a new breed of heroes: the public health engineers. Not only the Thames but the Seine, the Tiber and other major water thoroughfares in Europe had become increasingly foul with untreated effluent and rubbish. 'I counted two and seventy stenches – all well-defined – and several stinks', wrote Samuel Coleridge about the deplorable state of the Rhine running through Cologne. Safe water had to be provided to households by mains connection, and sewage led away underground to outlets closer to the sea. By the end of the century it was established in Europe and North America that civic authorities had a duty to provide citizens with household water and sewerage. Public Health Acts in Britain declared that a house without an adequate water supply was 'unfit for human habitation'.[2]

This enthusiasm for public health engineering had an important effect on average life expectancy in industrialized societies – so important that it became an axiom of sanitary wisdom that issues of disease control should be removed from the province of individual action into the realm of public administration. They were also to be paid for, or at least heavily subsidized, from the public purse.[3] The recognition that private companies would not or could not be expected to deliver services to all who needed them influenced the history of water supply, sewerage and drainage, not only in the industrialized countries, but elsewhere. In the common interest, clean water and drainage was a charge on the state.

As towns and cities mushroomed, the task of building and managing sewers and municipal water supplies was assigned to publicly funded authorities – in Djakarta, Buenos Aires, Delhi and Nairobi, as in Washington or Berlin. Until very recently, their dominant ideology was not a financial requirement to cover costs, nor an ecological need to conserve resources, but the engineering priority of meeting growing demand. The engineers were in charge of fulfilling the gospel of public health and citizens became their involuntary consumers.

Limits to growth

In the industrializing world of the late 19th century, the costs of pumps, pipes and drains were met – somewhat reluctantly, but met nonetheless – from the fruits of industrial progress. But the post-colonial developing world could not afford to provide more than a few urban consumers – let alone rural dwellers – with networks of household water connections, sewers and drains. In 1970, over 70 per cent of the world's population were still without safe water and 75 per cent without sanitation.[4]

In 1977, a UN Conference on Water was held at Mar del Plata, Argentina. A new generation of sanitary reformers prevailed on delegates to declare 1981-1990 the International Drinking Water Supply and Sanitation Decade (IDWSSD). Despite the monumental task confronting service deliverers, they set themselves an ambitious target: 'Water and sanitation for all'. The cost estimated at the start of the Decade for achieving full coverage with modern facilities was estimated at $3,000 billion. The figure is just one indicator of how over-ambitious they were.

The experience of sanitizing the industrialized world continued to color the international approach. The underlying ideology remained intact: access to water and sanitation services was a social right, justified on grounds of public health, to be provided principally at the public expense. Engineered supplies of safe water were axiomatic for gains in life expectancy and reduced rates of diarrhoeal disease. But if the ideology was intact, the existing model – water mains and sewer connections – was clearly inappropriate for those living in backward rural areas and urban shantytowns. The costs would be exorbitant and the technology inappropriate for flimsy or earthen dwellings. Yet these households contained the two-thirds of the global population relying on 'unimproved' water and toilet facilities. What could be done for them?

The hygiene-related disease toll

These households also contained those enduring most of the 80 per cent of sickness in the world that the World Health Organization (WHO) describes as water-related. These people are on the bottom rung of the social and economic ladder, their cash income is low, they have little education or access to health-care, their housing is poor, and their neighborhoods without amenities. Their efforts to maintain cleanliness and personal dignity in adverse circumstances are often extraordinary. But although they have strong views about the water they drink – what it looks like and how it tastes – their knowledge of hygiene is poor and they are vulnerable to disease.

Water-related diseases come in many forms [*see box*]. Parasites and germs may enter through the mouth or the skin. Others – scabies, trachoma – come from washing with dirty water. Water standing around in puddles provides the habitat in which some disease carriers – the malarial mosquito for example – breed.

Hygiene-related diseases

Water acts as a conduit for health, providing our means of washing, cleaning, bathing and laundry. But if it becomes polluted, water acts as a conduit for disease in food and beverages. Thus water's role in health is double-edged. But it is more accurate to regard disease as spread by poor hygiene than by water: hands washed in unsafe water are better than hands not washed at all.

Diarrhoeas

Many bacterial and viral varieties, including cholera, typhoid, salmonella, amoebiasis, shigellosis. Picked up from food, drinking water, hands and utensils (fecal-oral route); 4 billion cases of diarrhoea every year, and 2.2 million deaths mainly among small children. Washing hands can reduce diarrhoea by 35 per cent.

Skin and eye problems

These occur mainly in areas where water is short and people do not wash adequately. Trachoma causes blindness in 6-9 million people.

The most important infections in terms of their impact on human health are the diarrhoeas: cholera, typhoid, and other viral and bacteriological varieties. There are also some nasty exotic diseases, mostly dwindling and confined to locations in sub-Saharan Africa: river blindness, bilharzia, and guinea worm. These are all caused by parasites that live in water or in aquatic hosts and have debilitating consequences.

Reducing this disease toll, especially diarrhoeal disease, was the justification for Water Decade investments. Although it was also supposed to be a *Sanitation* Decade, drinking water supplies were more attractive and in demand. The Decade gave a much-needed lift to the provision of water to neglected rural communities: coverage rose to 50 per cent. Sanitation made far less progress: in fact, there were 300 million more people without sanitation at the end of the Decade than the beginning.

For those not on the urban planners' maps, low-cost technologies were developed by NGOs, with the help of backers such as UNICEF (the United Nations

---

Parasitic infections

Intestinal parasites e.g. roundworm and hookworm, affect around 10% of developing world populations, causing anaemia and malnutrition. They are passed in excreta and picked up on feet; wearing shoes is an important hygiene measure. Other parasites such as guinea worm and bilharzia live in aquatic hosts and are imbibed or enter the skin. Bilharzia (schistosomiasis) affects 200 million people.

Water-bred diseases

The malarial mosquito breeds on water; malaria causes 1-2 million deaths a year with 100 million cases at any given moment. Onchocerciasis (river blindness, 18 million cases of which 300,000 are blind) and dengue fever (30-60 million infections annually) are also carried by water breeding insects.

New Internationalist, No 354, March 2003; *Our Planet, Our Health*, Report of the WHO Commission on Health and Environment, 1992

Children's Fund), UNDP (the UN Development Programme); even the World Bank and a few commercial manufacturers took a hand. Although in these schemes the engineers still ruled supreme, at least their systems were simpler to build and maintain. Handpump-boreholes, rooftop rainwater harvesting, capped springs, covered wells, gravity-fed standpipes have since become staples in places where centralized mains are out of the question.

However, the new services did not have the impact on diarrhoeal incidence so confidently expected. This was a blow to the international protagonists of public health. What had gone wrong?

A matter of poor diagnosis

In the industrialized world, water and sanitation are two sides of one coin since the domestic water supply is used for drinking, washing and to flush toilets, removing waste at a stroke. So an emphasis on *water*-related diseases, obscuring the fact that their cause is really poor hygiene and lack of *sanitation*, makes little difference. But where there is no physical connection between water supplies and excreta disposal, sanitation has to be dealt with separately and this makes all the difference. What tends to happen is that sanitation and hygiene don't get dealt with at all.

For many reasons, including cultural distaste, lack of demand and corresponding lack of political impetus, the same investments have not been made in sanitation. Until sanitation and hygiene are given the prominence they deserve, diarrhoeal disease rates will not tumble. Meanwhile, the important water issue for disease control goes beyond safe water for *drinking* to the provision of plentiful water for *washing*, especially before preparing and eating food. Only when enough water is available to be used for hygiene, and is so used, does 'the fecal-oral route' of infection rupture.[5] If bacteria and parasites are constantly being picked up on hands and feet or washed into ponds where people bathe, a safe water supply cannot make the decisive difference to public health that it does in the fully engineered environment.

A safe water supply cannot fix health in such a setting without changed behavior. Safe, hygienic behavior depends on knowledge about germs and worms too tiny to be seen. And people already have their own theories of disease and its causes. 'Why are the flies sinners in our district?' asks a Nigerian villager. Proper information has to be sensitively put across.

Pushing supplies, ignoring demand

Lack of willingness to bite the bullet over the need for sanitation and hygiene education was not the only reason why many schemes performed poorly. Another problem stemmed from the long-held public health engineering article of faith that people needed water and sanitation services and must therefore be supplied with them whether they asked for them or not.

Where separate installations of pumps, boreholes, piped springs and tanks are concerned, it is essential for those benefiting from the improved services to play a significant part in managing and maintaining them. The services are not subject to distant, centralized command and their benefits cannot be imposed by engineers and bureaucrats. Since this had not been part of the industrialized world sanitary experience, this too was overlooked.

Stories about rural water schemes abound with woeful descriptions of installations which were inappropriate, unaffordable, broke down and were not repaired. Sanitation had its quota of similar disasters: smart new toilets constructed in people's courtyards, subsequently used as store houses or chicken coops. To begin with, the 'hardware' was blamed, and there were efforts to make pumps sturdier, cheaper and easier to repair. But the real problem was social, something that engineers found difficult to address. It was essential to consult people about the new facilities and where to put them. Many installations were not valued highly – they were in the wrong place or were not preferred to existing practice – and when they broke down no-one bothered to repair them.

The providers of the new facilities tended to ana-lyze their failures purely in terms of poor mainte-nance or local people's ignorance. But the main issue was simpler: lack of demand. Water supplies – and, in some places, sanitation – are wanted in many poorly-served areas. But not usually because the existing sup-ply is 'unsafe'. Sometimes, people know that seasonal epidemics of diarrhoea are connected to a particular water source, and then health is a motivation. More often demand is to do with distance and shortage. Women who rise at 4 am to queue in a distant river-bed for a basinful of murky water may be desperate for a pump near their homes. But where real demand does not exist and health benefits are not appreci-ated, people are often indifferent to 'improved' drinking sources.

It may seem curious that there are those among the 1.1 billion people without safe water who do not feel a demand for the services that the world wants them to have and believes they have a right to. But where these services do not meet their actual needs; are not within their capacity to operate; and where they have alternative sources however 'unsafe' they may be, 'improved' services fail. People vote with their feet and wallets to neglect them. A 1992 study in Orissa, India which looked into why a high proportion of new handpumps had fallen into disuse found that many yielding non-potable water according to chemical tests were in use, while others yielding good water were not. After analyzing 4,000 pumps, the researchers concluded that the only clear-cut reason why people used the new handpumps was if no other source was available or better located.[6] Meanwhile, where a pump is sorely needed, local householders will organize repairs and pay for them. Thus where the supply is valued, the installation is cherished even if the official maintenance system is poor.

The need to discuss in advance local demand and service preference is constantly repeated by organizations with experience in community water and sanitation. But despite endless international rhetoric about 'management by demand', many programs run by donors, governments and municipalities still fail to consult with prospective users. They ignore all their own precepts about stakeholder participation and decision-making at the local level, and carry on in full 'we know what is best for you' mode. For example, they insist on boreholes when people would prefer open wells. Not surprisingly, when they then ask the beneficiaries to pay for boreholes they never requested and don't particularly want, they baulk.

Impacts on disease

For a variety of reasons, therefore, the expansion of water supplies to new communities has not had the hoped-for impact on rates of diarrhoeal disease. Without enlarging the program horizon, it is impossible to achieve a public health engineering triumph in the non-industrialized world. Clean water may be essential for disease control, but a cause-and-effect relationship between the two is elusive. In fact, it is easier to deal with outbreaks of disease by medical technology such as cheap antidotes against malaria, parasites, and diarrhoeal dehydration. Only guinea worm has proved possible to eradicate by a specific focus on water quality. And even then, education to explain the guinea worm life cycle is also essential.

When the engineers took on their instrumental role in public health, and local bureaucrats and politicians promised to deliver free supplies, they removed from communities their sense of agency in managing local water resources. Their role, or a new version of it, has had to be restored. This is now happening because the authorities have had to accept that, with or without donor assistance, they will never be able to provide all their citizens with improved services without consumer contributions. This is not an argument in favor of the privatization of services, simply common sense. And when a community values an improved water supply, there is usually no problem over fees or maintenance contributions, as long as everything is fair, openly agreed and they can see where their payments are going.

Wise water practitioners today do not embark on any community water scheme without a joint planning exercise, a hygiene education campaign, and long-term commitment to system management. Gradually, consumers are being put back into the driving seat. If they need spare parts for their pump, or they have to call in the engineer, local water committees take the responsibility, with support from local authorities. These networks also offer promotional vehicles for cut-price toilets and environmental health. Not only does this mean services are more sustainable, but there is also more chance that their potential for hygienic living will be realized.

The sanitation challenge

Meanwhile, providing people with adequate sanitation grows daily more problematic. People in many rural areas prefer 'open defecation'; persuading them to dig a pit and install – at some expense – a slab, a toilet pan, and a superstructure is difficult unless they are already experiencing squalor, lack of privacy, and obvious disease. That is beginning to occur, especially in very fertile parts of Asia and Africa where the landscape is crowded.

Rising populations and denser settlement also place extreme strain on the earth's natural cleansing capacity. Water-borne sewerage is not only impracticable for low-income settlements, it is also wasteful of water and itself a cause of pollution. In fact, if the sanitary engineers were to start again from scratch today, the use of precious water as a human waste transport would probably be outlawed. Very small amounts of pathogenic wastes (surprisingly, only feces contain pathogens) along with perfectly sterile urine are added to very large quantities of water. In most parts of the world, this is deposited straight into rivers and the sea. In India, 80 per cent of the pollution load destroying the country's rivers is caused by untreated human waste.[7] In fully engineered environments, large sums of money are spent in removing the pathogens from this solution before sending the clean water into the rivers or back down the pipes again. Waterborne sewerage is therefore extremely water-inefficient, and for water-short and poverty-stricken areas, the most inappropriate system imaginable.

There are alternatives to sewers and septic tanks. The feature common to all the low-cost devices proposed since the days when Mahatma Gandhi was promoting sanitation is that they are dry and 'on-site'. They confine human waste to a dug pit or a brick box, where it can settle until the pathogens have died. Some are updated versions of the earth closet. In others a small amount of water is used for pour-flushing a polished cement pan. Above all, the on-site toilet should be congenial to use; otherwise people will continue 'open defecation' if they can.

A refinement of the non-waterborne approach is 'ecological sanitation'. Not only does the eco-toilet avoid flushing pathogens into waterways, it preserves the nutrients in human excreta (urine contains nitrogen, potassium and phosphates) and recycles them as fertilizer. The problem posed by eco-san in fully flushing societies is that it requires the use of a urine-diverting toilet. Eco-san has been well-received in China, and in some densely settled corners of Mexico, Uganda, South Africa and India. But enthusiasts insist on its usefulness everywhere. They believe that waterborne sanitation is becoming an extravagance the world cannot afford. And that losing a freely available fertilizer by discarding it as 'waste' is itself a waste of massive proportions.

People living in ever more crowded urban and rural settings are gradually coming round to the joys of the indoor toilet. But they need a certain sense of their privacy and status to find the investment worthwhile. A solid VIP (ventilated improved pit) toilet in Africa with a brick superstructure or a pour-flush toilet in Asia, can cost a household $250-500 – more than many spend on their entire living compound.

Playing catch-up

The combination of high costs and population growth is seen as the main reason why, despite all the new installations, the numbers of those without water and sanitation facilities remains roughly static. Growing

What you need, what you get

If all uses other than drinking, cooking and washing are ignored, and you forget about livestock and food plot cultivation, the figure of 30 liters of water a day is a good benchmark of need: 5 litres for consumption, 25 for hygiene. Many people have to make do on much less; while the average British citizen uses around 200 liters, and US, 500 liters.

Populations using the least amount of water

Country	Liters of water used per person per day
Gambia	4.5
Mali	8.0
Somalia	8.9
Mozambique	9.3
Uganda	9.3
Cambodia	9.5
Tanzania	10.1

Rob Bowden, *Water Supply: Our Impact on the Planet*, Hodder Wayland, 2002 needs consistently outstrip supply. The failings of the programs complicate the picture further. But there are other difficulties which cannot be laid at the door of public health engineers.

The reduction of flows in rivers and the depletion of shallow aquifers serving traditional sources such as wells, means that many of those who manage on only a few liters a head a day – compared to several hundred liters per head in sprinkler, fountain and swimming pool land – are finding even this small quantity increasingly difficult to come by. The pressures of agriculture and industry, and of expanding metropolitan cities, mean that water is becoming an increasingly scarce and inaccessible commodity.

The rural drinking supply water program in India, the largest in the world, illustrates the conundrum. Since the 1970s, the Government has been providing water to 'problem villages' – those whose supply is scant or distant, most in hard-rock areas – with UNICEF support. Here the dropping water table made it difficult to reach water in hard rock formations by traditional methods. So modern high-speed borehole drilling was adopted. Safe water was the emphasis, so groundwater became the eagerly promoted domestic water source. It is now used for drinking by 90 per cent of rural Indians.

Over time, a problem emerged. The number of 'problem villages' never seems to go down. Every time a new survey is conducted for the next five-year plan, however many boreholes have been installed since the last, the number of villages without a dependable supply shoots up again. NC Saxena, a former Secretary of Rural Development, commented: 'In our mathematics, 200,000 problem villages minus 200,000 problem villages is still 200,000 problem villages.'[8] Why? Because the water table is continuing to drop and many new wells run dry. India's groundwater aquifers are being heavily exploited for irrigation, and unless they can be adequately recharged, the rural water supply program is on a perpetual hiding to nowhere.

Over-extraction of groundwater across the border in Bangladesh has led to a different kind of drinking water crisis: contamination by arsenic. Scientists differ as to the precise set of geo-chemical circumstances which triggers the absorption of naturally occurring arsenic into groundwater: the phenomenon is new. But the lowering of the water table is regarded as central to the problem. By the mid-1990s, arsenic in groundwater was beginning to affect wide swathes of the countryside around the mouths of the Ganges River. In 1997, WHO declared that arsenic poisoning was a major public health problem in the area. In Bangladesh, 20 per cent of tested wells showed high levels of poisoning. Delays in addressing the arsenic problem in Bangladesh caused an international furor which has not yet fully died down.

India also suffers from arsenic in groundwater, but not yet to the same degree. Meanwhile, the presence of excess fluoride in groundwater is now also causing alarm. This too is a problem of higher concentration of the chemical in groundwater, associated with over-extraction of aquifers. Like arsenic, fluoride is taste-less, colorless and odorless. Ingested over a long period it does not kill, but it affects multiple tissues, organs and systems of the body and results in a crippling condition. Fluorosis affects no less than 62 million Indian people,[9] and around 70 million in northern China.

Rural people facing water stresses in the sources on which they depend are not confined to Asia. Population and livestock pressures, exacerbated by climate change, are prompting environmental degradation and water loss in many parts of Africa. In northern Namibia, seasonal water-holes used to last all the year round. Now they vanish within a few short months. Here, the groundwater is saline, so villages are desperate for piped supplies. Tanzania, whose arid north-west borders Lake Victoria, faces similar difficulties. It has recently been decided to build a 105-mile (168-kilometer) pipeline from the lake to supply 400,000 people with water for domestic use and livestock,[10] even though this flouts a treaty under which Egypt can veto any project affecting the flow of the Nile. Water for many of the world's under-served people is becoming something of far more serious consequence than possible ill-health.

And if drinking water catch-up is becoming more difficult and complex in the countryside, in towns and cities it is worse.

Thirsty towns

There are now about as many people living in towns and cities as in the countryside. Urban populations are growing at a spectacular rate, mostly in unplanned settlements where amenities are non-existent. Although the typical slum population in Southern cities is one-third to one-half, these inhabitants often do not feature in urban planning because they are not supposed to be there. Utilities are already overstretched providing services in the better parts of town.

Their problems are acute. Many cities have already exhausted nearby surface and underground sources. Amman, Delhi, Santiago and Mexico City are among those pumping water from increasing distances and up increasing heights. Some of Mexico City's supply has to be pumped from 300 kilometers away. Its pipes are in such poor shape that 40 per cent of the water leaks into the ground; and since the city is sitting on what used to be a marsh, it is sinking at a rate of 50 centimeters a year. Coastal cities such as Jakarta and Bangkok have so depleted their aquifers that seawater intrudes for many kilometers. In China, water needed by rural communities has been diverted to Beijing to stave off crisis; the water table below the capital has dropped 37 meters over the past four decades.

According to the World Bank, every time a new

Wastewater goes untreated

Only a small fraction of the wastewater collected in sewerage systems in developing countries is treated and disposed of properly. Most is discharged straight into rivers, lakes and seas, thereby endangering health.

Percentage of wastewater treated in selected capital cities (1998 or latest data)

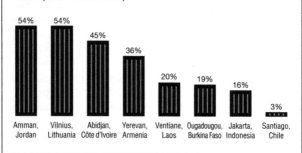

54%	54%	45%	36%	20%	19%	16%	3%
Amman, Jordan	Vilnius, Lithuania	Abidjan, Côte d'Ivoire	Yerevan, Armenia	Ventiane, Laos	Ougadougou, Burkina Faso	Jakarta, Indonesia	Santiago, Chile

engineering scheme replenishes a typical urban supply from farther away, the unit cost of raw water doubles.[11] Meanwhile the wastage from existing systems is inordinately high: in Manila, 58 per cent of water channelled expensively into city pipes is unaccounted for – it simply disappears. The record in such cities as Cairo, Mumbai and Nairobi is not much better. In Latin America generally, a recent review found that 40 per cent of the urban water supply vanished unaccounted and unpaid for. As the quantity of available water dwindles, disruptions – even sabotage – of existing systems become acute.

Cities on every continent are turning off the taps more frequently in the summer in an effort to conserve their supply. The increasing insanitariness of towns and cities is posing new challenges to environmental health. In 1990 cholera reappeared in Peru, having been absent from Latin America for nearly a century.[12] The epidemic spread to 16 other countries causing nearly 400,000 cases, and has since remained endemic on the continent. There have been other alarm calls. In 1994, 500,000 inhabitants of the Indian city of Surat – one-quarter of its population – fled from an epidemic of pneumonic plague. The disease had spread from the rotting carcasses of livestock washed up in the slums after heavy monsoon rains.

The call for reform

Playing catch-up with water and sanitation, let alone facing the growing problems of extra demand, higher costs and heavier pollution, requires radical new approaches. But what should they be?

When the problem of water scarcity and resource degradation came to the fore in the early 1990s, radical new approaches were indeed introduced. They followed a course predictable within the prevailing ideology of the day. If water was such a valuable and dwindling commodity, it should carry a price-tag and be subject to laws of supply and demand. Efficiency and conservation would then prevail. These ideas gained currency in the run-up to the first Earth Summit in 1992, striking at previously accepted principles.

The first of these was that water is so essential for survival that access to it has commonly been treated as a matter of public right. The second principle, established by the sanitary revolution of the 19th century, was that engineered water and sanitation services should be heavily subsidized in the interest of public health. This idea could be described as the way the earlier principle had been taken forward into the modern, industrial lifestyle era. Now a new principle was proposed. To overcome wastefulness, water should be charged for at an 'economic' price. This would mean that the state should no longer subsidize water supplies generously, if at all. The new regime would require new management. Since public utilities were making such a poor fist of providing essential services and recovering their costs, the private water industry would be invited in. The implications of the new approach were profound, as we shall see.

Before proceeding, one thing needs to be under-lined. Using a disease control rationale to squander water or provide useless services is no help at all in resolving the world water crisis. In the poor areas of the South, the industrialized public health mind-set has not always served the water-short effectively. People want a year-round source of water near their homes because they need water to live. Without a reli-able and plentiful source of water they cannot attain any decent quality of life or household productivity: there is an important economic element not captured either by 'safety' or 'convenience'. In fact, the per-ception which understands water poverty purely as an issue of drinking and washing is very distorting. In most traditional lifestyles, water's domestic and liveli-hood uses cannot easily be divided.

Water's importance in supporting rural livelihoods, including agriculture and the food supply, is where we'll go next.

1 *Global water supply and sanitation assessment report 2000,* WHO/UNICEF. **2** Colin Ward, *Reflected in Water: A Crisis of Social Responsibility,* Cassell, London, 1997. **3** SE Finer, *The Life and Times of Sir Edwin Chadwick,* Methuen, London, 1971. **4** *Towards the targets: An overview of progress in the first five years of IDWSSD,* WHO, April 1988. **5** John Thompson et al, *Drawers of Water II,* IIED, London, 2001. **6** Maggie Black, *A Matter of Life and Health,* OUP and UNICEF, New Delhi, 2004 (forthcoming). **7** Manoj Nadkarni, 'Drowning in human excreta', *Down to Earth,* Vol 10 No 19, 28 February 2002, CSE, Delhi. **8** Anil Agarwal, *Drought? Try capturing the rain.* Briefing paper. CSE, Delhi, 2001. **9** A.K. Susheela, 'Fluorosis management programme in India', *Current Science,* Vol 77 No 10, 25 November 1999. **10** Jeevan Vasagar, 'Storms lie ahead over future of Nile', *The Guardian,* 13 February 2004. **11** *The World Development Report 1992: Development and the Environment,* World Bank. **12** *Global water supply and sanitation assessment report 2000,* WHO/UNICEF.

3 Water to eat

The amount of water used in agriculture is far higher than that used for any other purpose. In many countries with little or erratic rain, irrigation is vital for food production. But the heavily engineered disruption of rivers which has characterized the development of irrigated agriculture in the past 75 years has produced a long catalogue of ills: displacement, water-logging, salinity, over-dependence on water-guzzling crops, and economic inefficiency. Is there no better way of combining land and water use for food?

ALTHOUGH THE GLOBAL water crisis is most often projected in terms of drinking water, water is as vital for our food supply – and needed in far greater quantity. When drought strikes in Ethiopia, Niger or Rajasthan, even if the well does not run dry, the land becomes barren, livestock die, and when their store runs out, people are forced to migrate in search of food. Those living off the natural environment – the world's 1.4 billion small farmers and their families – may encounter this at first hand. For the rest of us, a temporary hike in the price of certain food items or a hosepipe ban which stops us watering our vegetable patch is the closest we get to a life-and-death encounter with water as food.

The proportion of water used for drinking and domestic purposes is very small. In global terms, it amounts to only 10 per cent of water extractions and in many developing countries, much less: in India, for example, only 5 per cent. So although people have become more extravagant in their home water use, domestic users are not chiefly responsible for water stress even where showers, toilets and washing-machines are standard amenities. Nor are industrial users, although their share is growing, today reaching 20 per cent. The remaining 70 per cent is used for growing crops on irrigated lands. In the driest tropical areas, the figure rises to 90 per cent.[1]

The take-off of water from rivers, lakes, and underground aquifers for agriculture has a very long history. Ancient civilizations in Egypt, China, Jordan and the Indus Valley all depended on irrigation to maintain their food supply; ruins of irrigation canals have been found in Mesopotamia (modern Iraq) dating back 8,000 years. In the 19th century, irrigation was crucial in opening up the American West to farming prosperity. Waterworks have helped fruit and vegetable gardens to flourish in unpromising dryland areas of Spain, California, Australia and Israel. But traditionally, this was not the norm. Except in climates where rain was seasonal or almost non-existent, agriculture was rain-fed until a century or so ago.

As populations grew and spilled out into more marginal lands, so did the need for food and the impulse to green the land by moving water about the countryside. The US Bureau of Reclamation was set up in 1902 – 'reclamation' meant putting arid land under irrigation – to construct projects financed by selling government land and, later, by selling water and electricity. Its glory years began in 1931 with the blasting for the Hoover Dam on the Colorado River. The 221 meter Hoover, 85 meters higher than any other dam then built, was regarded as a technological feat on a par with any in history and inspired an almost spiritual awe. But other monsters followed. The Grand Coulee on the Columbia in Washington State, an irrigation dam 1,500 meters long and 168 meters high, was described by an enthusiastic US Senator as 'the biggest thing on earth'.

These dams ushered in a modern era of hydraulic might on a par with that of the Nebuchadnezzars of the ancient world but deploying vastly superior technology. The vagaries of nature regarding rainfall and river courses for the purposes of watering dry lands could now be permanently corrected. For several decades this creed was not in doubt, as political leaders and their corps of engineers took on as a public duty the serious business of building large water-retaining structures, re-designing rivers, creating vast reservoirs, and generally demonstrating that hydrologists and engineers could improve on nature's arrangement of water in the landscape for agricultural productivity and other noble purposes.

This technocratic adventure was pursued with enthusiasm by many leaders of newly independent countries. The majority of large dams were constructed between the 1950s and 1980s. Only in the last two decades of the 20th century did it come under strenuous attack by environmental and social activists. Nobody questions that millions of farming livelihoods, not to mention national granaries, depend on managing and manipulating water flows for crop growth. But does the provision of water for the replenishment of food baskets, especially among those whose plates are usually half-empty, really require an ever more destructive makeover of river valleys and watersheds and the uncontrolled extraction of non-renewable sources underground?

The drive to irrigate

Pakistan is one country where the drive to irrigate on a massive scale is understandable. Dry and largely rainless, Pakistan could not survive without the waters of the Indus. This 2,880 kilometer river carrying twice the flood of the Nile flows out of the glaciers around Mount Kailas in Tibet. Fed by snow melt and Himalayan mountain streams, the Indus descends into Kashmir and then into the plains below, where it is joined by five other rivers. Together, over the centuries, these have deposited alluvial silt a mile deep over the palm of fertile Punjab.

The British built a complex system of canals all over the Punjab. At independence in 1947, the partition of the subcontinent left Kashmir in Indian hands, and with it the headworks. Very quickly, tensions developed. When India temporarily halted water flows to carry out maintenance, riots erupted in Pakistan. The two countries, constantly at the brink of war, had to find a way to share the waters of the Indus basin. This led to the 1960 Indus Water Treaty, allocating the waters of the three western rivers to Pakistan, and the three eastern to India. India allowed a 10-year grace period during which Pakistan undertook construction of a network of dams and canals. Supported by the World Bank, the Pakistanis pulled off a staggering engineering feat: two of the largest dams in the world, 19 barrages, 57,600 km of canals, 1.6 million km of watercourses and ditches – the largest contiguous irrigation system on earth.[2]

In 1951 when this program was first mooted, what was then West Pakistan had 34 million inhabitants; 50 years later, there were 152 million. The 35 million acres of land opened up to irrigation have helped to sustain that population. And this is only one among hundreds of large-scale irrigation projects based on dams, reservoirs and canal works constructed all over the industrializing world. A long line of post-colonial rulers fell under their spell. Nothing seemed to become a great leader so well as a shining massif of concrete stretching away into the distance or towering into the sky. The vast majority of the world's estimated 45,000 large dams (those over 15 meters high, or having a reservoir area of over 3 million cubic meters) were built after 1960, mostly in Asia. Half are entirely or mainly for irrigation, and in Asia, the proportion is 65 per cent.[3]

This explosion of hydraulic construction increased the amount of cultivable land placed under irrigation five-fold in the last century. The major acceleration came alongside 'green revolution' agricultural techniques and cheap electricity for water-pumping. Some of that cheap electricity is produced by hydropower: many dams are dual purpose. The high-yielding plant varieties used to expand the world's food basket have

Water for food, water for power

Some large dams are constructed partly or entirely for power generation (the latter is most common in Europe and South America). Hydropower is an important and clean source of renewable energy. However, opposition to large dams on social and environmental grounds does not favor this production route, especially as evaporation from large reservoirs is extremely wasteful of water. Small-scale hydropower plants generating less than 10 megawatts, enough to power a small town, are ideal for remote areas where connection to the national power grid is unrealistic. At community level, hydropower holds out many possibilities, especially where electric pumps are used for irrigation and the cost of power is key to agricultural production.

Distribution of large dams by region and purpose

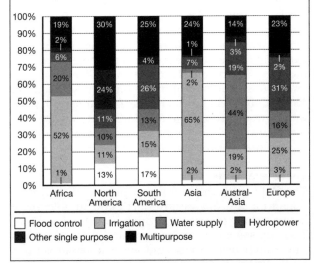

Dams and Development, the Report of the World Commission on Dams, 2000

to be combined with fertilizers and pest control, and rely on irrigation at just the right moment to promote optimum growth. Since 1960, engineered withdrawals of water from rivers, lakes and aquifers for agriculture have rapidly increased. And as the world's population rose inexorably, irrigation became the cornerstone of global food security.

The proportion of sunny, warm and fertile lands converted into important crop-producing areas by irrigation does not sound enormous: around 17 per cent of cultivated areas. But their contribution to the food basket is disproportionately high, according to the UN's Food and Agricultural Organization (FAO). In developing countries, irrigated land produces 40 per cent of all crops and 60 per cent of cereals: these provide 56 per cent of calories consumed worldwide.[4] If irrigation were suddenly to end, yields in critical grain-growing areas of northern China, northwest India, and the western Great Plains in the US would drop by between one-third and a half.[5]

Indeed, according to the world's experts, so essential is the role of irrigation in the expansion of the global harvest and therefore in the world's ability to feed itself that the amount of water required is expected to rise by 14 per cent in the next 30 years as the area of irrigated land expands. The important question is what kind of irrigation from what kind of water sources will be employed? The large-scale kind, with its vast constructions and attraction for corporate investment and under-the-counter deals; or something more sensible and human-scaled?

There are limits

The FAO tends to paint a rosy picture of irrigation, suggesting a strong positive link between investment in irrigation, poverty alleviation and food security. This is a statement of the totals-and-averages variety, with which many social and environmental activists would disagree. Any idea of expanding vistas of land irrigated by extensive canal networks, dams and reservoirs, and of ever larger water withdrawals from dwindling lakes, rivers and aquifers, begs many questions.

Where water is scarce, competition between users is bound to arise. In India's Kaveri basin, for example, there is an annual stand-off during the dry season between the states of Karnataka and Tamil Nadu about the behavior of upstream Karnataka in retaining more than its fair share of Kaveri waters for its struggling farmers. Sit-ins, riots and attacks on vehicles near the state boundary can be anticipated in years when the monsoon is less than abundant. In 1991, water riots in the two states displaced 100,000 people.[6] Violent incidents over water withdrawals for farming have become part of the seasonal political landscape all across the subcontinent: barely a day passes without a clutch of newspaper items about farmers at boiling point over water for their crops.

Then there is a small matter of expense. The amount of land being brought under large-scale irrigation worldwide began to slow in the late 1970s and has not since noticeably picked up. This is partly because the most cultivable lands and the most accessible sources of water have already been used. The costs of expanding irrigation through large construction projects have become much higher and it is now much more difficult to procure finance. The capital costs per irrigated hectare run over $4,000 for large projects in China, India, Pakistan and Indonesia; in the Middle East, they reach $5,000; and in Africa, where roads and infrastructure are lacking and the irrigable land not very large, they reach $18,000 or even more. External backers are put off, especially as the economic outcome of irrigation schemes has rarely met expectations. Unfortunately, however exorbitant these amounts become, they fail to discourage determined water moguls who procure amounts from the public or private purse with scant reference to cost-effectiveness or accountability.

No scheme could be more grandiose than Colonel Qadhafi of Libya's $32 billion 'Great Man-made River' which will lead water from an aquifer under the Sahara through 3,500 kilometers of pipelines to irrigate his country. The promise of abundance is

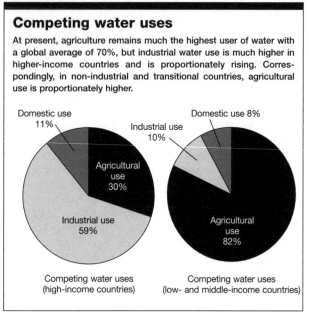

Competing water uses

At present, agriculture remains much the highest user of water with a global average of 70%, but industrial water use is much higher in higher-income countries and is proportionately rising. Correspondingly, in non-industrial and transitional countries, agricultural use is proportionately higher.

Domestic use 11%

Industrial use 10%

Agricultural use 30%

Industrial use 59%

Competing water uses (high-income countries)

Domestic use 8%

Agricultural use 82%

Competing water uses (low- and middle-income countries)

ephemeral at best: the aquifer is non-renewable as no rain lands in the Sahara and its contents are expected to last Libya between 15 and 50 years at most. After that the Sahara may subside or the emptying aquifer suck in water from the Nile, leading to drastic consequences. Never mind that it would cost far less to import food. Even extremely arid countries – Namibia is another example – insist for geopolitical reasons that they should be self-sufficient in food production no matter what the water costs.

Another serious limitation is the difficulty of running large-scale canal and watercourse systems and keeping them in repair, especially in resource-strapped countries. In Pakistan, one-quarter of the crash-built system of 40 years ago is now water-logged. Part of the problem is lack of drainage, which should have been built at the time, and is now very expensive to install on a steadily dilapidating system. Where too much water is used or canals are unlined, the water table rises. Too much water chokes off oxygen from the root systems of plants, interferes with the rotting of organic materials, reduces nitrogen and encourages the accumulation of toxins in the ground.

The rising water table brings another hanger-on: salt. As surface water evaporates, a deathly white crust is left behind. Estimates suggests that 47.7 million hectares, over one-fifth of the world's irrigated land, is 'salt-affected'.[7] Around 2 million hectares of soil become so saline every year that they have to be abandoned. In extreme cases of over-irrigation, whole water systems have dried up: Lake Chad, bordered by Sudan, Chad, Nigeria and Cameroon, has shrunk by more than 90 per cent since its feeder rivers were over-taxed. In both Egypt and Pakistan, saline and water-logged soils mean that the crop is 30 per cent less than it should be. Drip irrigation – used by Israel and in parts of the US – is far more efficient. But to transfer to this technology is too expensive for many irrigation-bound economies to bear.

And then there is groundwater. Large dams provide water to around 30 per cent of the world's irrigated area. Water from underground supplies most of the rest. Few countries have a full picture of their groundwater reserves, some of which lie so deep that they are inaccessible for all practical purposes. Meanwhile, pumps powered by cheap diesel and electricity are being used to extract groundwater for irrigation at a far faster rate than rainfall can renew it. In the US, more than 4 million hectares – roughly a fifth of the country's irrigated land – is watered by pumping in excess of aquifer recharge. The most famous aquifer in the world – the Ogallala, which stretches from the Texas panhandle to South Dakota and contains 20 per cent more water than Lake Huron in the Great Lakes – is being depleted by irrigation pumps at a rate 14 times faster than nature can restore it.[8]

In India the picture is even more dramatic. Courtesy of cheap electric power and 20 million mechanized pumps affordable even by modest farming families, around half of the area under irrigation is supplied by groundwater extracted faster than the water table can recoup. The inevitable accompaniment has been the drying up of shallower aquifers, a reduction of water flows in rivers, and the progressive deepening of wells. Where a borehole depth of ten meters used to suffice for a plentiful supply, 80 meters may now be required.[9] In the worst-affected 'dark zone' areas, village wells have to be re-filled by tanker.

Irrigation works for whom?

Although the importance of irrigation to the world's food supply is frequently underlined, those who tend to gain from large-scale irrigation works are not the world's hungry. On the contrary, whatever their stated intentions, the benefits of these schemes tend to end up in the pockets of larger landowners, while poorer families lose out. This is a variation on a familiar theme: put a complex piece of machinery between the world's least well-off and their food supply, and you push the price of food beyond their reach. Major irrigation works and the rest of the industrial agriculture package that goes with them may boost yields per hectare, fill national granaries and produce an exportable surplus; but they also permanently destroy millions of livelihood systems.

The story begins when entire populations are displaced. The World Commission on Dams estimated that between 40 and 80 million people have been displaced by large dams, although the actual figure may be much higher. Most receive pitiful amounts of compensation and end up impoverished. A disproportionate number are from indigenous groups wedded to their ancestral lands, but without means to defend their rights to them. Even where the displaced are promised new land to replace that submerged or crushed under millions of tons of concrete, this is rarely in the new irrigation 'command area'. Resettlement may be to an alien environment. Thousands of families displaced by the Sardar Sarovar dam in India's Narmada Valley are expected to move to a different state among different people – as if Scots displaced by a reservoir were forced to move to France.

Inland fishing is also affected by dams and irrigation works. Most of this catch is consumed locally or marketed domestically and is an important source of food and livelihood for poor people. The blocking of rivers by dams has a damaging effect on wild fishing by altering up- and downstream flows. Migratory fish cannot pass, nor can other organisms critical to maintaining aquatic ecosystems and food chains. Large dams have played an important role in the rapid loss of freshwater species, around one-third of which are now classified as extinct, endangered or vulnerable.

In Asia, aquaculture is often introduced into new reservoir areas. But the commercialization of what used to be hunted as a 'free' resource may put both catching and eating fish beyond the means of those who most need to do so. It is the usual story: those who live off a natural resource base adequately or well in their own terms are evicted to make way for those who may be able to gain a commercial profit in the short term, whatever the real costs over the long term in human and environmental terms.

Other strange things happen at hydraulic construction sites. Land values around new canals immediately shoot up and poorer cultivators are squeezed out to make way for sugar-cane or cotton plantations operated by commercial farmers with an eye to cash crops and export markets. These can afford the equipment, chemicals, labor and know-how. They also have the clout to fix officials and make sure the canal goes just where they want it to. In this way of things, previously independent and self-sufficient small-holders are reduced to the status of landless laborers, sometimes paid a ludicrously low wage to concrete over the very plot of ancestral land that they have lost.

Even where poor farmers are given land in a large scheme, they often end up in debt. This happened along the Indira Gandhi Canal in the Thar Desert of Rajasthan. Within 20 years, two-fifths of the new settlers lost their land to money-lenders – for whom they now work in debt bondage – because they could not cope with the new type of farming economy. Meanwhile the Canal enjoys all the usual desert irrigation problems: uneven water distribution, poor drainage, an epidemic of water weed, salinity, water-logging, mosquitoes and malaria. Despite its record, senior water officials in Rajasthan still think that vast canals will be their salvation. They talk in terms of volumes – crop volumes, water volumes, land-under-irrigation volumes – and assume that every technical problem merely awaits a technical fix.

Water and inadequate livelihoods

Water is fundamental to the livelihoods of those whose subsistence depends on the natural environment. Yet when it comes to describing 'poverty' and 'the world's poor', the link between water and basic livelihoods is rarely mentioned. We are frequently told that 1.1 billion people are without safe drinking water and 2.4 billion without sanitation – the only watery mentions in the Millennial Goals for reducing poverty by 2015. There are other ways of being water-poor. What about those people whose lives are constantly on the edge because their water resources, along with other parts of the natural resource base, are being steadily depleted – often in the name of 'development' for others better-off than they are?

One study by the International Food and Agricultural Fund (IFAD) suggested that out of 4 billion people in 114 low-income countries, more than 2.5 billion lived in rural areas; of these more than half lived on highly degraded soil and around 1 billion lived below the poverty line.[10] Are the 1 billion people living in degraded environments at the peril of drought and water shortage the same 1 billion plus people who, according to World Bank figures, are living on less than $1 a day? Do they have any connection with the 1.1 billion without reliable drinking water? There must be some overlap and if their situation is to be tackled, their water resources access and use will have to be looked at afresh. Water to drink is not going to solve all their problems.

Water is critical to the interaction between such people and their livelihood base. For the majority of poor people, this consists of lands, forests, rivers, coastal waters, and small-scale entrepreneurial activity based on natural products. While water is today described as an 'economic good' which should be charged for at a realistic price, its role in traditional economic systems has been ignored. So too are the systems themselves: many development commentators take as a given that semi-subsistence smallholder agriculture is inefficient and has to be replaced. Inefficient for whom and in consuming what resources? As far as inefficiency with water is concerned, it is not the subsistence farmers who are at fault.

Since water's exploitation by societies operating according to pre-industrial norms does not have a price-tag attached to it, its role in household productivity and food security has not been noticed. Although its value may be well-understood to a local community, the traditional water source has been treated as economically invisible and under vacant possession. This allows planners to behave towards these resources as if they were dealing with uncapitalized state assets – claiming them for their own version of the 'common good'. When environmental changes are made which alter people's access to water, reduce its flows, or pollute it and thereby drive out a source of income such as fish, a vital economic prop is removed. When a stretch of untamed river is impounded behind a dam and becomes a reservoir, the rights of local inhabitants to fish in the river or to quarry its banks may be lost. Its contents – waters, fish, sand, shorelines – become subject to licensed exploitation: fee-payers or bribe-bearers only are welcome.

Water-efficient crops

Critics of the Green Revolution, notably Indian environmentalist Vandana Shiva, deplore the way in which the role of water in traditional food production systems has been ignored by advocates of industrial agriculture. Food cultures evolved in response to the water possibilities in a vicinity. In the wetter parts of Asia, rice cultivation emerged. In arid and semi-arid areas, wheat, barley, corn, sorghum and millet were the staples. In highland regions, pseudo-cereals such as buckwheat were grown; in deserts, people survived by livestock-herding.

This diversity and its adaptability to local soils and climates is being lost, laments Shiva, without reference to the subtleties of water's role in crop productivity. Monoculture, which suits agro-business interests, is the order of the day. Sorghum and millet which use less water to produce the equivalent in nutrition are dismissed in favor of high-yielding rice and wheat. Most water-consumptive of all are livestock, especially cattle. The area of land under cultivation of pulses and coarse cereals has rapidly declined in many parts of the world, despite their superior water conservation and cost-effective nutritional advantages. It is ironic that the advocates of water-at-a-realistic-price only discovered water's economic value in agriculture after they had pushed farmers out of water-efficient methods into water-profligate crop dependency. Also their talk of increasing 'crop per drop' is rarely to advocate a reversion to diversity and water-conscious cropping, but to advocate genetic modification in the classier cereals: rice and wheat.

Water use by crops

Different types of food require more or less water for cultivation. Pulses, roots and tubers use least; starchy cassava is known in Africa as the famine crop. Among cereals, rice uses more than twice as much water as wheat and maize; millet is better able to withstand drought than all of them. Sorghum, for the same amount of water, produces 4.5 times more protein, 4 times more minerals, and can yield 3 times more food than rice. Ironically, cattle are herded where land supports no cultivation at all: whatever their water consumption needs, they are the ultimate desert larder.

Product, per kilo	Water required in cubic meters
Beef	15
Lamb	10
Poultry	6
Palm oil	2
Cereals	1.5
Citrus fruit	1
Pulses, roots and tubers	1

UN World Water Development Report, 2003

The water-consumptive, agricultural export promotion strategy does even less for the wellbeing of those whose traditional systems are displaced. Along the Senegal Valley, where paddy-field irrigation was supposed to raise dietary levels, rice consumption among villagers rose; but consumption of traditional foods – millet, sorghum, maize, legumes and pulses – fell by 30-90 per cent. A study funded by the US Agency for International Development (USAID) in 1994 found that the nutritional status of the people in the Valley had declined. The people blamed the loss of their previous foodstuffs; these they could no longer grow because they had planted them seasonally as waters receded on flood plains destroyed by the scheme.[11] The story is familiar among 'beneficiaries' of similar projects in Sudan and Tanzania.

Thus the direct connections between the reduction of malnutrition – which affects around 750 million people worldwide – and the replacement of traditional farming systems by modern irrigated agriculture are elusive at best. To have enough to eat, a household needs either enough land to grow food or enough cash to buy it, and some reserves for difficult times. The state of the national granary is irrelevant for a family with no access to its contents. For too many marginal farmers, a combination of factors which includes the depletion of once reliable water sources is making their outlook ever more parlous. In this sense, water-poverty is one of the factors fuelling migration from the land by people desperately seeking a livelihood in town.

Harvesting water and food

Although sub-Saharan Africa is talked of as having unfulfilled irrigation potential, the likelihood of more massive irrigation projects on the model of the Nile Basin is fading. Existing schemes have proved problematic and prone to failure, their costs are exorbitant and conditions in most river basins are not conducive. However, many small-scale traditional methods have had considerable success. Since these are built and run by farmers themselves, they are often ignored in official government statistics. When FAO corrected its irrigation figures to include them, the irrigated area in Africa rose by 37 per cent.

Traditional methods include *dambo*-farming – the intensive cultivation of small garden plots in seasonally water-logged lands. *Dambos,* meaning 'valley-meadowlands', occupy around a tenth of central southern Africa, including parts of Angola, Malawi, Mozambique and Zimbabwe, and typically provide up to half of a *dambo* farmer's income.[12] In drought years they can be essential in seeing families through. Heavy machinery would probably damage the soil and water regimes of these wetlands, but with low technology they are environmentally sustainable. As in the case of other local water harvesting techniques used elsewhere on the continent for terracing contours, bunding soil, and holding rainfall in the ground, *dambo*-farming promotion could do much to boost food production. But unlike the subsidies and services which accompany grand irrigation schemes, there is little assistance in the form of credits, loans and technical advice to promote small-scale water conservation in farming. The vegetables produced on *dambos* are not seen as credit-worthy, and nor are the women who are frequently their cultivators.

Informal or small-scale systems can account for large areas of irrigation: in the Philippines, about half the irrigated area, and in Nepal, around three-quarters. In India, although traditional forms of water harvesting and 'tank' irrigation were pushed into obscurity and disrepair for decades by state provision of public works, there has recently been a determined renaissance in their favor. This has been pioneered by figures such as Anna Hazare, whose introduction of water harvesting techniques – percolation tanks, check dams, bunding of *nullahs* (streams), contour building, additional dug wells, aquifer recharge – in a drought-ridden and destitute village in Maharashtra transformed it into a thriving agricultural center.

Village water in Gujarat, India

Micro-watershed development in an Indian village has raised incomes and dramatically reduced migration from the land.

Thunthi Kankasiya village	before 1991	1999
Perennial drinking water wells	nil	23
River dams	nil	1
Months of water availability	4	12
Land under cultivation (hectares)	85	135
Number of crops per year	0-1	2-3
Agricultural production (quintal/hectare)	900	4,000
Seasonal migration rate	78%	5%
Period of migration (average, months)	10	2
Income per household (rupees per annum)	8,590	35,620

Reviving water wisdom

Since the 1980s, the work of Hazare and similar eccentrics in Gujarat, Madhya Pradesh and Rajasthan has begun to capture notice as a viable approach to local water and food security problems. In 1997, the Centre for Science and Environment (CSE) in New Delhi published a report: *Dying Wisdom: Rise, fall and potential of India's traditional water harvesting systems.* This represented the accumulation of 10 years of work, led by the late Anil Agarwal and Sunita Narain, two of India's most prominent environmental activists. It was originally inspired by the discovery in a remote corner of the Thar desert of unique traditional devices – *kunds* or carefully constructed saucer-shaped catchments – for collecting and storing rainwater. These were allowing people to meet their water needs even at the height of a devastating drought.[13]

Since then, CSE has campaigned with missionary zeal to promote the use of traditional water harvesting systems which performed so well throughout the subcontinent until colonial engineers and their Indian inheritors destroyed them or let them fall apart. The wealth of evidence CSE has collected of the effectiveness of many types of systems in meeting local needs is breathtaking. And the assertion that communities can, with a little investment, re-green their environment, replenish local aquifers, and transform their farming fortunes has attracted widespread notice, including from former President KR Narayanan and many other figures of national and international importance.

One of the striking attributes of these projects is that they combine the management of land and water within the local river valley or watershed. The replanting of trees on eroded slopes to enable rainfall to be retained in the soil rather than cascade away is an integral component. Since aquifers are recharged and water levels are thereby restored in boreholes and wells, there is also no artificial division between providing water for drinking and providing water for other livelihood purposes including the growing of food. This holistic view of humanity's many uses of water, and the attempt to manage the total land and water resource base in a mutually reinforcing manner, is the way things were done before governments invented 'sectors' – health, agriculture, industry, environment – and the modern focus on national and corporate balance sheets took over.

Today, in the interests of water conservation and efficiency, water policy experts call for 'integrated water resources management': this is the new holy grail in balancing uses and moderating fairly and cost-efficiently between them. Funnily enough, it's what traditional farming communities in water-short areas have been doing all along. Is it possible that, given a different outlook, the prospects for managing the world's water sustainably are really better than they look?

1 *Comprehensive assessment of the freshwater resources of the world*, World Meteorological Organization and Swedish Environment Institute, 1997. **2** Diane Raines Ward, *Water Wars: Drought, Flood, Folly, and the Politics of Thirst,* Riverhead Books, New York, 2002. **3** *Dams and Development: Report of the World Commission on Dams,* November 2000. **4** *Water for people, water for life,* UN World Water Development Report, UNESCO, 2003 **5** Sandra Postel, *Last Oasis: Facing Water Scarcity*, Worldwatch Institute and WW Norton, New York, 1992. **6** Vandana Shiva, *Water Wars: Privatization, pollution and profit*, Pluto Press, London 2002. **7** Sandra Postel, *Pillar of Sand: Can the Irrigation Miracle Last?* WW Norton, New York 1999. **8** Maude Barlow and Tony Clarke, *Blue Gold: The Battle against Corporate Theft of the World's Water,* Earthscan, 2002. **9** UNICEF Water Section New Delhi, India, *A Time Line*, 1998. **10** *The State of World Rural Poverty,* IFAD and IT Publications, 1992. **11** Patrick McCully, *Silenced Rivers: The Ecology and Politics of Large Dams,* Zed Books,London, 1996. **12** Eds. Munyaradzi Chenje, Phyllis Johnson, *Water for Southern Africa,* A Report by SADC, IUCN, SARDC, Zimbabwe, 1996. **13** Introduction to *Dying Wisdom* by Anil Agarwal and Sunita Narain, CSE, New Delhi, 1997.

4 Water at a price

Public utilities have a poor record in the developing world for delivering water supplies and sanitation. In the early 1990s, the idea that 'water is an economic good' was co-opted by international exponents of the neoliberal agenda and their corporate allies. Privatization of utilities – and contracts to the burgeoning water industry – was supposed to make water services efficient and expand them to the poor. The strategy failed.

'GOD PROVIDED THE water, but not the pipes.' This is how an official at Suez, the world's biggest water company, explains the need for his industry.

If we want taps, faucets and toilets which flush, then an industry there has to be. Water is heavy, it only flows downhill, it finds any tiny outlet for escape, and its transport is expensive. But for over a century (as Chapter 2 explored), the view was that the operations which laid on the pipes and taps should be in the public domain and that services should be heavily subsidized. Water was special: its life force should not be a source of profiteering. In the late 19th century, Britain accepted a duty to provide every household with a supply regardless of ability to pay. In the late 20th century, everything changed. In 1994, 12,500 households in England and Wales had their water disconnected because they had not paid their bills.[1]

Pioneers in sanitary reform, in 1989 the British became the pioneers in market reform regarding the stewardship of those pipes which, indeed, God did not provide. Instead, Mammon took a hand. Prime Minister Margaret Thatcher privatized the water industry, floating all ten English and Welsh water utilities on the stock market. This was done mainly for ideological reasons as part of her religious revival of the market. But the way it was done – with debts forgiven, a lavish dowry to compensate for lack of recent investment and freedom to hike up prices – led to fantastic profits and huge 'performance-related' pay-outs to the new directors. Between 1989 and 1995, customer charges jumped 106 per cent, and profit margins for the companies increased by 692 per cent.[2] To understandable public shock, water had become a commercial product under monopoly control, whose profits now lined shareholders' pockets.

In the 1990s, the privatization of water utilities else-where in the world advanced under the banner of market reforms along with the urge to commodify public goods, reduce 'big' government, cancel subsidies and social programs, and thereby sort out inefficiency. It was all part of the neoliberal agenda about how countries should run their economies and govern themselves effectively. At the same time, another trend in international thinking emerged: enhanced recognition that the natural environment was under pressure from development processes. And freshwater was among those components of the global commons needing protection.

The idea that freshwater was a precious resource which was being wastefully squandered by bad policies and worse practice came at just the moment when market ideology ruled supreme. A conjuncture between the two ideas was inevitable. Agenda 21 had made the sensible observation that 'water is an economic good'. But this became a banner under which subsidies for domestic supplies should cease in order to reduce water waste. Water must be charged for at the price it cost to capture, treat and deliver. Public utilities that failed to cover their costs could not be excused because they were operating for the common good. Since in many places poorer citizens had no service at all while the better-off received heavily subsidized supplies, that case was anyway hard to plead. Utilities had proved themselves inefficient or corrupt; so the remedy was for them to give way to, or become, private companies.

Britain's full-scale privatization of its water-related

assets did not provide the template for the water privatization policies advocated over the next decade. But it certainly laid the ideological groundwork. In defiance of the lessons of the past, private rather than public provision was now expected to meet poorer humanity's outstanding needs for taps, pumps and spigots. And if people could not pay, then what?

A business like any other

Delivering water to households and removing wastewater from them, declared a recent *Economist* survey,[3] is best done by treating water 'as a business like any other'. This is also the conclusion of a March 2003 report on the financing of water schemes by a panel set up by the World Water Council (WWC), chaired by Michel Camdessus, a former managing director of the International Monetary Fund (IMF). A panel full of bankers and international corporate representatives was likely to think along those lines. But should consumers of water services be treated like consumers of other goods and utilities, some of which may be basic but not fundamental to human existence?

The idea that access to something so essential as water should be left to the mercy of the market has been greeted with outrage. But does this mean there is no role for private services, or that pipes and pumps should all be free? That would be equally absurd.

Many ancient towns and cities had – some still have – informal systems of water provision where people pay. In Bhuj, Gujarat, the shops in the market alleyways buy their water daily by the *debbie* or can. In Merca, Somalia, and other towns in the dusty Horn of Africa, water collection and distribution still keeps a host of donkey cart owners and their boys in business. As the city fabric becomes more sophisticated, modern works take over. But major problems arise with an extremely rapid pace of urbanization and with the high proportion of citizens living in slum and squatter areas. As new shanty-towns spring up, public service utilities originally designed to provide services for a few hundred thousand people have been over-whelmed by populations soaring into the millions. Those in the poorer end of town shift as they can.

The parlous state of water and sanitation provision in fast-growing cities has been obscured by the way the authorities describe 'service coverage'. In Africa, 85 per cent of the urban population is reported as having access to improved water services and in Asia and Latin America, the proportion is 93 per cent.[4] These figures are based on dubious notions, such as that a standpipe every 200 meters equals 'universal access' to water.[5] While in the countryside this might be acceptable, in the crowded slum with hundreds of dwellings dependent on one tap, it is not. Studies show that despite the claims of many cities in Asia and Africa to provide improved services to over 90 per cent of inhabitants, between 30-50 per cent have very inadequate services. Less than half the inhabitants of cities in Africa have water piped to their homes.

Since local utilities cannot provide a reasonable or in some cases even a minimal service – if squatters are 'illegal' the authorities usually provide no service at all – people in the poorer parts of town are often exploited by private operators. No doubt the growth of public utilities in 19th century Europe and North America came about in part from the need to deal with this kind of unregulated and exploitative prac-tice. In the late 20th century, when urbanization was taking an even more headlong course in the South, the lessons drawn were different.

The conclusion should have been that the political economy of water and sanitation services in the indus-trialized world was unsuitable and unaffordable for the straining urban fabric in Africa, Asia and Latin America, and that locally adapted options should be encouraged to emerge. There were many issues sur-rounding how to build and pay for vital services in problematic municipal environments. But they all

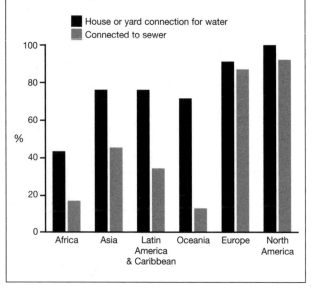

Households in major cities connected to taps and sewers

The figures are based on information provided by 116 cities. The picture in towns which are not major cities can be worse since they do not attract similar political attention or investment.

■ House or yard connection for water
■ Connected to sewer

WHO and UNICEF, 2000

became submerged by belief in the prevailing dogma about markets and privatization.

The water giants enter the picture

In the early 1990s, the World Bank undertook a review of its experience in investing in water systems over two decades. Out of 120 projects, the authorities were performing well in only four countries. The long list of inefficiencies included failure to supply connections, charge for them and keep them in repair; only 35 per cent of costs were typically being recovered.[6] Meanwhile, other studies showed that poor families in cities such as Lima and Port-au-Prince spent up to 20 per cent of their incomes on water from vendors and paid far more per bucketful than middle-class customers with household connections. These twin discoveries combined to suggest that market efficiency would turn things around in everybody's favor.

The World Bank therefore began to extol the virtues of private sector participation in water and sanitation services. Various models of 'public-private partnerships' were suggested, from service contracts for operations and maintenance, to BOT – build, operate, transfer – arrangements whereby assets, financial investment and commercial risk were also divested.[7] A green light to private sector participation, using the expertise available from the industrialized world, was signalled as the way forward. Loan policies of the development banks actively encouraged municipal authorities to take this route. And the international water companies stepped up enthusiastically.

The leading water service companies were not in the US – where 90 per cent of utilities are publicly owned – but in France. Here, companies such as Générale des Eaux and Lyonnaise des Eaux had grown up over decades on the strength of concessions from municipal authorities to deliver the service, conduct repairs and collect the revenues, keeping surpluses as profits. Since 1990, when the privatization boom began, these two companies, reincarnated as Véolia (previously Vivendi) and Suez (within Ondeo), have become huge multinational corporations. Between them they own or have a controlling interest in water companies in 130 countries; Suez claims 115 million customers and Véolia, 110 million. Behind come other power and utility corporations such as Bechtel, Enron (until it collapsed) and Britain's Thames Water (now owned by the German company RWE), competing for lucrative contracts.

Between 1990 and 1997, there were nearly 100

Water giants

Although there was a huge and highly contentious push in the late 1990s to encourage the water industry to invest in service delivery in the South, increasing private sector involvement has produced neither the scale nor the benefits anticipated. Only around 5% of the world's customers for water are currently served by formal commercial providers. After many setbacks, that proportion is unlikely to rise any time soon.

Sales of groups and water divisions, 2001

Water division	Water sales (euros m)	Worldwide customers (millions)
Ondeo	10,088	115
Véolia (formerly Vivendi)	13,640	110
Thames	2,746	37
SAUR	2,494	36
Anglian	936	5
Cascal	181	6.7
IWL	100	10

H₂O, Guardian special supplement, August 2003

cases of non-local private companies taking over water supplies management in developing countries; in the preceding six years there had been only eight.[8] Corporate involvement was heavily boosted when insistence on privatization of public utilities became a part of 'structural adjustment' packages. A recent review of IMF loan policies in 40 countries revealed that, during 2000, agreements in 12 included conditions imposing water privatization or full cost recovery. Examples included debt relief for Tanzania, which required the assignment of the country's water authorities to private management; similar conditions were imposed on Rwanda and Niger. Honduras was required to approve a law for the privatization of water and sewerage systems. Nicaragua had to increase its water tariffs by 1.5 per cent a month on a continuous basis and offer concessions for private management of services.

The overt courtship of transnational water corporations by international bodies claiming to pursue development and combat poverty caused consternation among social activists. Suspicion was fuelled by statements such as that appearing in *Fortune* magazine in May 2000: 'Water promises to be to the 21st century what oil was to the 20th century: the precious commodity that determines the wealth of nations.' The industry's membership in international quangos, notably the World Water Council which laid on World Water Forums in Marrakech (1998), The Hague (2000) and Kyoto (2003), reinforced the impression of an onslaught against the global commons undertaken by an international agency-corporation cartel set on milking profits from a vulnerable resource at the expense of vulnerable people. At the 2000 Water Forum, the impassioned protests of activists nearly derailed the entire event.

The enthusiasts for privatization think their intentions have been wilfully misunderstood. They worry that industry will mop up available water cheaply leaving domestic customers in serious difficulty unless sensible tariffs are introduced. If water delivery is more efficient and better-off customers properly charged, then it will be possible to spread the service to poorer neighborhoods and subsidize lower-income customers. And if those who have more water than they need – certain farmers for example – can sell it to others who are running short – such as municipal authorities – mutual benefits could accrue. The promotion of these price-driven efficiencies and 'water markets' naturally requires that the necessary regulatory regime is in place. It also requires that complex institutional machinery has all the competencies and structures to implement it. Hence the policy experts arrive at their usual refrain that everything will work out beautifully in the presence of 'good governance'.

Water at a price

The trouble is that 'good governance' in the form of efficient, democratic, transparent and accountable institutions, and regulations which not only exist in a statute book but are actually carried out, is not on offer in the majority of African, Asian and American countries. Certainly, as water companies have learnt to their cost, it is difficult to operate where municipal governance does not conform to their exacting standards. And this is only one reason why the vision of water services expanding to the poor courtesy of Véola and Suez has spectacularly failed.

A good story turns bad

To begin with, the experience with private companies seemed positive, even for poorer urban areas. A successful example of public-private partnership lauded by the Asian Development Bank was in Manila, a city notorious for its wasteful water record. In 1997, a Suez-owned consortium called Maynilad won a 25-year concession for water and wastewater services for the West Zone of Metro Manila.

Hundreds of 'blighted areas' containing 35 per cent of households had been obliged to buy highly-priced water from private vendors. The company determined to offer everyone a piped supply. People had a bad experience of public faucets and chose individual connections. A 'Water for the Community' project built a supply-line above ground, against the walls of narrow alleyways. Batteries of meters were installed, and plastic pipe connections led off to every house that wanted one. The system allowed 58,000 additional household connections to be metered and charged for, in places where dwellings were rudimentary and even 'illegal'.[9] The tariffs, to begin with, were lower than those charged by the public utility.

This experience shows that where imagination is used and socio-economic conditions are right, expansion of services by private operators to the urban poor in a way that serves their interest is practicable. Here, the company even took the trouble to discuss with communities what kind of services they wanted. But such honeymoons do not last. When Maynilad tried to hike its tariffs to cover these and other investments, the municipality protested. In 2002 Maynilad pulled out, citing debts and broken agreements.

The idea of charging the full-cost price for water is not feasible, leaving aside the ideological pros and cons. It will fall foul of local politics for a start: those who have the clout will prevent prices rising to such levels – this is the reason why in many cities tariffs remain 'improperly' low. This is a political problem and cannot be solved by commercial diktat. Municipal authority may be sympathetic to the need to raise tariffs – when the service improves. But prices cannot be hiked beyond a certain point; and sometimes the service never improves enough to justify tariffs higher by several multiples. The sums simply do not add up.

Water companies tied into the international market, no matter their social responsibility and green pretensions, have to make money for their shareholders. Indeed, that is why they were so enthusiastic in the privatization boom of the 1990s to do so in emerging markets. But with the best will in the world, which is more than many would grant them, it has not been easy. Companies, having burnt their fingers in Manila, Jakarta or elsewhere are now thinking twice.

Latin American experiments

Buenos Aires provides another example. The water concession was awarded to Aguas Argentinas, a subsidiary of Suez, under the terms of a Presidential decree. In 1993, the company began repairing the ageing system, and within six years had increased the proportion of the population receiving water from 70 per cent to 82 per cent. Although tariffs went up, the new consumers at the lower end paid one-tenth for their piped supply what they had paid to vendors in the past. The profits to the company were considerable: 21 per cent of revenues in 1997.

Buenos Aires was held up for many years by the World Bank as its copy-book example of water privatization success. But in 2002, when Argentina had a catastrophic devaluation, the economics of the operation fell apart. The Government refused to let Aguas Argentinas raise prices to offset the devaluation and Suez pulled out. A promising marriage ended in a rancorous dispute in the arbitration courts.

Staunch believers in market orthodoxy say that an excellent scheme was ruined because there was no devaluation back-stop facility to pay out to the company in a financial crisis. Others draw a different conclusion: that it is not desirable to let an internationally-owned enterprise take on the responsibility for a service which, first and foremost, has to serve the public interest of the residents of Buenos Aires rather than its dollar-bound shareholders.

As for the notion that squeezing the consumer will help conserve the resource, this too is an illusion. Such corporations do not have as a priority the need to conserve resources. Maximizing profits means encouraging increased consumption. Extending services to thousands or even millions of households whose consumption is tiny – and who will keep it that way because they cannot afford to do otherwise – is not a cost-effective proposition.

The most well-known experiment in bringing in an external consortium to lay on water supplies was in the Bolivian city of Cochabamba. Here, the World Bank had pressured for privatization of water supplies, but actually refused support to the Mayor's pet scheme: a dam and tunnel they saw as too costly. The Government went ahead anyway and gave the contract to a company called Aguas del Tunari, owned by the US giant Bechtel. When water tariffs were doubled to pay for the investment, and people who had built their own rooftop tanks and pipelines began to be charged for using them, protests erupted. After mass demonstrations early in 2000, the Government gave in, the project was scrapped and the contract torn up. Bechtel has since taken out a suit against Bolivia for $25 million in damages.

Beating the retreat

When Aguas del Tunari fled from Cochabamba, a victory was claimed on behalf of anti-privatization forces. That a group consisting of labor activists, NGOs, professionals, and local politicians had seen off an effort led by a US corporation to commodify the rain was a triumph. The protest movement convincingly claimed the moral high ground and the story has become a modern parable of people's resistance to the corporate theft of the commons.

Stalwart supporters of the water business approach insist that examples of failed private enterprise do not have generic faults. There are explanations: the Mayor of Cochabamba's poor judgement, poor local 'governance', bad pricing decisions, Argentina's devaluation. Many of the companies themselves disagree. Suez has announced that it will reduce its activities in developing countries by a third. The managing director of another company, SAUR, told the water division of the World Bank that the private sector cannot deliver for the poor: 'The scale of the need far outreaches the financial and risk-taking capacities of the private sector.'[10]

The retreating forces recognize that to turn the provision of water into a going commercial concern with returns for international shareholders in an environment whose economy cannot adequately support such a venture is bound to fail. There may be efficiencies that commercial expertise can help introduce for some parts of the system. But not for low-income communities. David Hall of the University of Greenwich puts the problem thus: 'The fundamental issue is that the poor are not profitable.'

Not profitable for connections laid on by Suez and

Bechtel, at least. But there is a viable water services economy in which the poor take part, which seems to have escaped the experts' attention.

Invisible service providers

To the experts, the informal water services currently used by the poor appear to be invisible. Not only do the commercial companies fail to make their own water economics pan out, but in the process of finding out that the poor are not a viable prospect, they destroy the ways by which they have up to now been served. Instead of regulating local entrepreneurial activity and providing a structure in which it could improve, existing providers tend to be undermined. This was an important finding of case studies in Dar es Salaam and elsewhere conducted by WaterAid and partners in 2003.[11]

Where it is recognized, this kind of service is demonized because it has been responsible for the relatively high prices poor families in the slums have been paying. These vendors, scavengers and cess-pit emptiers have been excoriated, while being conveniently cited as evidence that the poor can and will pay for supplies. No doubt some informal providers merit their exploitative reputation. But the picture presented is very one-sided, and the fact that the services they render are privately operated never seems to earn them any brownie points at all.

The assumption that the substitution of these unrecognized services by companies answering to Suez or Thames will improve matters for poor urban dwellers over the long term is simplistic. If the World Bank and other external agencies and investors want services to be privately supplied, the 'enabling environment' is not one which gives temporary favors and monopoly guarantees to corporate outsiders. It is one in which local initiatives have some chance of success, within a regulatory and institutional framework run by a public authority on an ethical,

The poor pay more for water

So what's new? It is common for people with fewer resources, less choice, and less influence to pay more per unit for food and power. So it is not surprising that those who buy water from a vendor pay more than those with house connections. This has been used as an argument in favor of utility privatization. Since a utility with a private sector partnership has not only to recover costs but make profits, it cannot offer services to the poor more cheaply than local vendors. Since water services have been heavily subsidized as a norm everywhere until recently, the figures chiefly illustrate how amenities are skewed in favor of the better-off. The question that arises from them is how to regulate the vendors and expand competitive service coverage. Neither publicly owned nor privately owned utilities are likely to reach these populations, and subsidies of some kind – as accepted in the past – will definitely be required.

Price for 1 cubic meter of water, selected cities 1997

household connection		informal vendor
$0.08	Dhaka, Bangladesh	$0.42
$0.16	Jakarta, Indonesia	$0.31
$0.14	Karachi, Pakistan	$0.81
$0.11	Manila, Philippines	$4.74
$0.09	Phnom Penh, Cambodia	$1.64
$0.04	Ulan Bator, Mongolia	$1.51

Atlas of Water, Earthscan, 2004

not-for-profit service delivery basis.

In water supply and sanitation, as in irrigation, a kind of apartheid seems to operate. The small-scale, the local, the informal entrepreneurial or the technologically modest activity that has served people up to now and offers a solution to their needs is usually discounted by the large players. For all the billions of dollars the World Bank has invested in water supply and sanitation – about 14 per cent of its budget since its inception – most of the benefits have accrued to transnational construction companies and the largest local industries. Probably less than 1 per cent has gone into small-scale ventures which do something for the seriously water and sanitation deprived.

If they don't pave the way who will follow? Where are the companies, doing 'a business like any other', who are prepared to invest in VIP latrines, roof-top water tanks, handpumps, lined wells, eco-sanitation, small-bore pipes and sewers and the other technologies which are the only practical ones for nearly half the people in the world?

The alternatives

The reality is that the provision of clean water and improved sanitation for poorer communities has almost exclusively been done by NGOs and some local authorities supported by a handful of big donors. There have been some striking successes based on community management of low-cost systems, where self-administered tariffs pay for some or most of the costs. NGOs may not be part of the profit-making private sector, but they often apply 'demand management' principles more effectively than government and utility service-providers. Sadly, these programs operate below the horizon at which many larger players notice anything going on.

WaterAid, the NGO set up and partially funded by the water industry in England and Wales, has tried hard to find ways to merge thinking across the private and public sectors. WaterAid points to a central principle of any successful attempt to establish services that people will pay for and use: the need for social mobilization and community consultation. How often is this prerequisite repeated in international policy documents, but when it comes to activity on the ground, community participation in any meaningful form is seen as a burden and ignored.

WaterAid has conducted a special research exercise into private sector participation (PSP) in water and sanitation, and reached the conclusion that PSP cannot help resolve the many shortcomings of current systems. The same conclusion has been reached by the authoritative environmental organization, the London-based International Institute for Environment and Development (IIED). Weak government capacity to deliver services may be a problem, but handing everything over to the private sector is not the answer. IIED believes that the whole private sector-public sector polemic is a distraction which allows all the important things which need to happen to extend services to poor people to be ignored.[12] In the meantime, projects repeatedly exhibit all the previous failings. When the scheme collapses and the PSP partner pulls out, the last situation is worse than the first.

Initiatives to expand services to the poor, whether undertaken by the private service companies or publicly managed utilities, need to understand that interaction with communities is essential. Only then will services respond to demand, thereby creating a match between a service and its market. Surprisingly this simple truth eludes most PSP initiatives. Similarly, the idea of making an inventory of existing informal service providers and seeing how they could be incorporated into a new scheme, does not arise. The model typically favored by privatized utilities and their international backers is the industrialized world public health engineering model transported lock, stock and barrel to much poorer cities, with minor adaptations and little reference to its technical, financial or managerial suitability for the majority population.

Good private practice
It is possible to widen private sector participation to include the use of artisans and small local contractors. In Uganda, for example, local councils identify projects, and once they have been approved at district level, companies or NGOs are selected via a tendering system to undertake construction. The community provides 10 per cent of the costs, but the rest is paid

for by the Government and WaterAid. A covered well with a pump costs around $1,400 and is within the community's competence to manage and repair. UNICEF supports a similar scheme in Rajasthan, India. Here, the contracted NGO is responsible for making a village inventory of water-points, undertaking health education, agreeing with the community on the location of new facilities towards which they make a 10 per cent capital contribution, and establishing committees to run each installation.

Most of these programs have no problem with concepts which have been wrongly co-opted to promote the idea that 'water is a business like any other'. Villagers and slum-dwellers definitely regard water as an 'economic good' – life cannot function without it. Most can and do pay for services; but they are not willing to pay for services which have been designed by outsiders and foisted onto them. Local control and accountability of service providers is crucial. If this is 'demand management', then communities are for it. The difference is that these services – handpumps, rooftop reservoirs, standpipes at regular and short distances, dry sanitation – are affordable and appropriate. Local masons and pump mechanics can and do make a business out of them. But not Bechtel or Suez.

Such programs cannot easily be designed on scales that allow millions of new household connections or thousands of new pumps. Creating community structures, with women's representation, to manage them is not a simple process. So much easier, as we do in the industrialized world, to have a connection, pay the bills and let the company get on with it. But the water and sanitation model developed in the industrialized world cannot be transposed to developing countries as if it is a universal fix. When it did not transpose well in the form of public utilities, there was no reason why sending in the transnationals would radically change matters for the better. The mystery is why the experts thought it would.

Reaching the millennial goals

At the 2002 Earth Summit in Johannesburg goals were set for water and sanitation: reducing by half the numbers of those without services by 2015. This means – if you accept the official figures – that by 2015, there will be only 550 million people without clean water and a mere 1.2 billion without sanitation. Whenever the possibility of attaining these goals is discussed, the question is almost always posed in terms of finance. What would it cost? Where will the money come from? Rarely does anyone describe what kind of systems are envisaged or how all the problems of providing services to the poor will suddenly dissolve.

Finding the necessary financial resources was the genesis of the World Water Council panel set up under Michel Camdessus. If finance is posed as key, private sector involvement is naturally seen as vital; it is the only way to multiply resources significantly. Much has been made of this argument by the World Bank – indeed the tiny proportion of the world's population served with water by private corporations was one reason it pursued this agenda so aggressively. The Camdessus panel followed the same line, suggesting how the private sector could win more contracts in developing countries, while insuring themselves against the sagas that have marred their prospects up to now.

The panel suggested that the present $75-80 billion invested in water in developing countries a year would have to rise to $180 billion, mostly coming from the international water industry. It is difficult to understand the mathematics of this proposal. If the record of the private sector continues to be as disastrous at meeting the needs of the poor as it has been so far, this sum will be needed just for damages in the arbitration courts. IIED's researchers are convinced that whatever the useful roles private companies can play in service provision, privatization 'should not be promoted internationally as if it provides the key to achieving the water and sanitation targets within the Millennium Development Goals'. WaterAid is even more categorical: 'It is very unlikely that the multinational private sector is going to play any significant role in reaching the Millennium Development Goals.'

WaterAid itself puts the figure of meeting the Millennial Goals much lower, commensurate with its emphasis on low-cost and appropriate systems: an extra $35 billion a year. But until it is possible, via public utilities, private partnerships or by any other route, to invest more of whatever resources are available into low-tech, locally affordable and manageable services, all talk of investment volumes is spurious. It is not the money which represents the key problem in meeting the needs of the billions of people without water and sanitation. It is the whole misguided approach.

There are now hopeful signs that the World Bank has recognized that the involvement of the commercial private sector is not the panacea they once implied. But in the meantime, the doctrine of making a business out of water by granting concessions to commercial interests has rapidly spread, with many unfortunate consequences. What began as private sector involvement in water and sanitation *services* has led to private sector exploitation of the resource itself – a much more sinister development.

1 Colin Ward, *Reflected in Water: A Crisis of Social Responsibility,* Cassell, London, 1997. **2** Maude Barlow and Tony Clarke, *Blue Gold: The Battle against Corporate Theft of the World's Water*, Earthscan, London 2002. **3** John Peet, Survey: 'Water', *The Economist*, 19 July 2003. **4** *Global Water and Sanitation Assessment Report 2000,* WHO and UNICEF. **5** *Water and Sanitation in the World's Cities,* UN-Habitat, Earthscan, London 2003. **6** *World Development Report 1992,* The World Bank, Washington. **7** Toolkit: *Selecting an Option for Private Sector Participation*, The World Bank, 1997. **8** Constance Elizabeth Hunt, *Thirsty Planet: Strategies for Sustainable Water Management,* Zed Books, London, 2004. **9** *Poverty Reduction and Integrated Water Resources Management,* Global Water Partnership, Stockholm, 2003. **10** Charlotte Denny, 'The poor are not profitable and foreign firms are pulling the plug', *The Guardian*, 23 August 2003. **11** *New roles, new rules: does PSP benefit the poor? Case studies in private sector participation in water and sanitation in ten countries*, Water Aid and

Tearfund, 2003. **12** Jessica Budds, Gordon McGranahan, *Privatization and the Provision of Urban Water and Sanitation in Africa, Asia and Latin America*, Human Settlements discussion paper, IIED, London, 2003.

5 A resource under threat

The freshwater ecosystem is at the hub of all ecosystems. In recent decades severe damage has been done to rivers, coasts, deltas, lakes and marshes. Restoring their ecological health is difficult and expensive. The people who suffer most are those whose culture and livelihood system were constructed from, and still depend on, the natural environment around them. Both the quantity and quality of freshwater are under threat from industrializing humanity and its current tendency to heat up the world.

THE IDEA OF the natural world as a web of interacting ecological systems in which freshwater played a pivotal role only emerged in mid-20th century. Before that, conservationist angst in North America and Europe about human intrusion on areas of great natural beauty was regarded as an élitist and romantic preoccupation. But then it became clear that more was at stake than a patrician enthusiasm for unspoilt vistas, nature trails and wildlife observation.

The story of the Everglades is instructive. Early last century, at the instigation of speculators, farmers and buccaneering politicians, the marshes of northern Florida began to be diked and drained by the US Army Corps of Engineers. The idea was to transform a useless 'wet desert' into a sculpted farming landscape of easy plenty. Gradually, some conservationist critics began to realize that the obstruction and channelling of water on a massive scale was irreversibly altering what today would be called the integrity of Florida's freshwater ecosystem.

In 1947, pressure led to the creation of the Everglades National Park, a sanctuary for the natural swamps and wildlife to the south of the already reclaimed area. However, the changes in the upstream water landscape continued to have a drastic effect. The fight to mend the damage inflicted on one

of the world's richest sub-tropical wetland wilder-
nesses has continued ever since. Billions of gallons of
freshwater continue to be pumped into the Atlantic
while the Everglades dies of thirst. 'This is a man-
made drought, the result of hubris, faulty engineering, and an almost wilful misunderstanding of
nature,' writes Diane Raines Ward.[1]

At enormous price, we have learned that reorganiz-
ing an environment to suit human purpose – whether
desert, valley, forest or plain – brings with it an environmental makeover which causes dangerous upset
over the longer term. Wetlands are today recognized
not as useless swamps and 'wet deserts', but as complex and fragile ecosystems destroyed at humanity's
peril. Draining the Everglades not only wiped out
amphibian habitats and reduced bird populations by
over 90 per cent, it also reduced rainfall, since the
sheets of water which used to evaporate into the
atmosphere are now being pumped into the sea.
Today, 'reclamation' in the Everglades has gone into
reverse. Instead of sending in the Army Corps with
dredgers to cut the bends out of rivers and build
canals and levees, they are now being used – at similarly vast expense – to put the bends in again, dis-
mantle choked canals, flatten levees, open sluices and
let the waters spread.

Nothing could better illustrate the change of view
about the value of water in the landscape. It is now
known that if the water in a given ecosystem is altered
significantly in quantity or quality, the hydrological
cycle is affected and the ecosystem itself will change,
taking away 'free' or natural support for the health of
all life cycles. In the 'economic good' language of
today, these characteristics of freshwater are described
as 'in-stream values'. They include the support of all
freshwater aquatic life – fish, plants and crustaceans –
and water's irreplaceable role in earth's inbuilt self-
cleansing system. If this is overloaded by water's loss
or pollution, the food chain is threatened, livelihoods

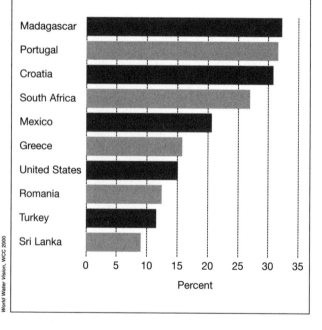

Life under threat

Biodiversity losses have only been partly detected and measured: just a few larger organisms are monitored. Worldwide, around 20% of freshwater fish are vulnerable, endangered or extinct; 20% of threatened insects have aquatic larval stages; 57% of freshwater dolphins and 70% of otters are vulnerable or endangered. In the US, 75% of freshwater molluscs are rare or imperilled. Half the world's wetlands have been lost, and eco-system integrity has declined in 35 million km of rivers following construction of dams.

Freshwater fish species threatened, selected countries

Madagascar
Portugal
Croatia
South Africa
Mexico
Greece
United States
Romania
Turkey
Sri Lanka

0   5   10   15   20   25   30   35
Percent

World Water Vision, WCC 2000

affected and biodiversity lost.

Maintaining water's in-stream values is therefore not merely a matter of aesthetics, recreation or marginal concern for creatures with exotic names. Losses may deplete natural systems in ways that imperil future generations' health and food supply. The cost of attempts to restore natural ecological values are very high – witness the billions of dollars spent in the Everglades and on programs to clean up major rivers such as the Thames and the Rhine – and most poorer countries say they cannot afford them. So despite the acknowledged need to make special allowances for nature's water requirements, the battle is far from won. In many places it has barely begun.

Still, a tendency exists to see environmental protection as a luxury which immediate needs for survival and development eclipse. But water is finite and more of it cannot simply be 'developed'. Without protecting the resource, life itself will be in trouble.

Disrupting the flows

The massive disruption to freshwater flows has come about partly because of humanity's overspill into 'virgin' areas and partly because of hydraulic construction on free-flowing rivers. Around 50 per cent of the world's wetlands have already been drained and the flows of around 60 per cent of the world's major rivers interrupted by human-built structures. Inland lakes and water bodies have shrunk and suffered habitat degradation. Some rivers are so reduced by take-off that they can no longer be relied upon to reach the sea. Both the Colorado River and the Yellow River belong in this category.

According to one recent analysis, human impacts on the hydrological environment have increased nine-fold since 1950. Only a portion of this impact stems directly from the rapidly rising withdrawals of water for irrigation, industries and domestic consumption. Most of it stems from human manipulation of natural flow patterns through the construction of dams, reservoirs and dikes. According to Sandra Postel and Brian Richter: 'Species that evolved over the millennia within earth's aquatic ecosystems are now reeling from these human-induced impacts. We have cast them into a race for survival for which they are not evolutionarily prepared. By virtue of our domination, we have become their stewards.'[2]

When you read about the needs of natural systems for water, and see them justified on recreational and wildlife habitat grounds, there is a moment when you wonder how strong, really, is the biodiversity case for water resource protection. Can these needs be as vital to life on the planet as the lack of safe drinking water faced by over one billion people and the scarcity confronted by at least as many trying to grow food and maintain their livelihood systems? But after another moment of reflection, you realise that the issues are all part and parcel of each other. The damaged wildlife habitat is the damaged resource base on which many people's entire life system depends.

Mangroves and coastal fringes

Take the destruction of coastal fringes. Large areas of mangrove forest have disappeared from African shorelines and the coasts surrounding the Bay of Bengal. For centuries, these provided a buffer against storm surges and tidal waves. Mangroves also treat effluent, absorbing excess nutrients such as nitrates and phosphates, and preventing contamination of shore waters. Because mangrove trees have special features, such as aerial and salt-filtering roots and salt-excreting leaves, they can survive in saline wetlands. And the ecosystems of which they are the linchpin have long provided a livelihood – food, medicine, fuel, construction materials – for coastal peoples.

Recently, unregulated expansion of aquaculture – shrimps for export – in Orissa and Andhra Pradesh in India and in the Sundarbans of Bangladesh, have led to widespread destruction of coastal mangrove forests and their replacement with shrimp ponds. With this has come the ruin of fishing communities' livelihoods. It is no coincidence that these areas of Andhra

Pradesh are one of the main areas on the subcontinent where human traffickers operate, 'buying' young girls for domestic service and the sex trade, and sending them into slavery far away. The ecological destruction which has led to this human misery is seen by environmentalist Vandana Shiva as an outcome of trade liberalization which elevates lucrative commercial contracts for shrimp above the well-being of indigenous communities.

Shrimps are not the only problem. The Sundarbans, the largest mangrove forest in the world, is an area of low-lying island mudbanks in Bangladesh's south-eastern coastal belt. A decrease in the volume of freshwater flowing into the Bay of Bengal due to dams and barrages on the Ganges is causing such salinity in the forest, as well as silting its network of canals, that the Sundarbans ecosystem is collapsing. The unique Sundari trees in large parts of the forest are salt-diseased and fish stocks decimated. Bangladesh's Environment Minister recently suggested that the decline of sweet water and the intrusion of salt in the coastal areas will lead to major agricultural loss. 'If this continues,' he said, 'we might have to consider relocating 15 million people.'[3]

Water-related disasters

The loss of mangrove swamps has made coastal peoples more vulnerable to flood from the sea. Experts suggest that the 1999 super-cyclone which devastated Orissa in Eastern India and caused immense land, crop and property losses as well as taking around 20,000 human lives, would have been far less devastating if the mangroves had not been decimated.

Other disaster-related problems start at rivers' headwaters. Many occur in the top levels of catchments, even before streams finish falling off steep hillsides and rushing to the valleys below. Forest canopies on the slopes should act as natural dams, holding rainfall back in their roots and lodging it in the mushy forest floor, from which it is slowly released into the atmosphere through green leaves (evapotranspiration), or into aquifers and streams. Deforestation, which has overtaken so many upland areas in the Himalayas, the African highlands, the Andes, Central America and elsewhere, prevents accumulation of water in the soil. Rainfall runs off the bare earth, swiftly adding volume to the fast-moving upper reaches of rivers. Flood risk in the valleys below is raised, and accelerated erosion on the upper slopes increases their vulnerability to drought.

The numbers of water-related natural disasters – especially flood disasters – have increased in recent years. Upstream ecosystem change is one reason. Another is that more people now live in floodplains. Where rivers naturally burst their banks to accommodate large seasonal variations in the speed and depth of flows, protection from rising waters becomes ever more problematic.

In China's Yangtze basin, farmers have turned floodplains and flood diversion zones into farmland. Land reclamation has shrunk major lakes or caused them to disappear. Local embankments have been built inside dikes, hemming in the river and raising its level. In 1998, also a year of drastic flood in north-east India and Nepal, the Yangtze flood crests were the highest ever recorded. Millions of soldiers and citizens mobilized to shore up dikes, dynamite levees and save downstream cities. The floods affected 16 million people, killed 3,656 and degraded 8.5 million hectares of land.[4] These were not the only major floods in China during the 1990s – just the worst.

Between 1991 and 2000, the number of people affected by natural disasters rose from 147 million per year to 211 million per year. Of these 2,557 natural disasters, 90 per cent were water-related, with floods representing 50 per cent. The recorded economic losses from these catastrophes grew from $30 billion in 1990 to $70 billion in 1999, and these figures are

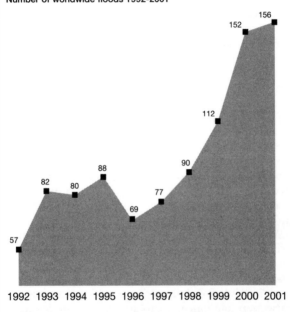

Floods and other watery disasters

Water in one way or another is behind most natural disasters, including droughts, famines and landslides where deforestation is a contributing factor.

Rising floods

Number of worldwide floods 1992-2001

57 82 80 88 69 77 90 112 152 156

1992 1993 1994 1995 1996 1997 1998 1999 2000 2001

Atlas of Water, Earthscan, 2004

thought to understate the true scale of loss by as much as 100 per cent.[5] They do not show the disruption to livelihood systems for so many of the people affected, many of whom live in marginal rural economies, and their assets – however essential to them – are typically disregarded in economic assessment.

The rising natural disaster trend is attributed partly to the damage inflicted on river basin systems. But climate change is also a part of the picture. In 2000,

A resource under threat

Mozambique experienced its worst floods in 150 years, with 700 people drowned and 550,000 displaced. In August 2002, two weeks' torrential rain filled the Danube to record heights and caused the worst floods in Europe in 200 years. Such crises confirm in the public mind that freak meteorological events are now becoming normal and – at least while they are in the headlines – bring home the terrifying reality of climate change.

Climate change and water resources

Floods and droughts are not the only water-related impacts to be expected from climate change. Most commonly cited is the rise in temperature, the melting of the polar ice-caps and the rise in sea level leading to the inundation of low-lying islands and coastal areas. Less obvious but potentially more serious, according to global water pundit Peter Gleick, are the changes to be expected in rainfall patterns and behaviours, in evaporation, and in amounts of run-off and moisture in the soil at different times of year: all the important variables for managing water in its multifarious uses.

What does this mean in human terms? It is most acutely felt in places where life is constructed around a local natural resource base operating within previously settled – if occasionally volatile – parameters. Traditional socio-economic systems – not to mention Darwinian biological adaptations – stem from the behavior of temperature and rainfall from season to season and their interaction with the earth. Climate change means that the calculations on which whole cultures are based are no longer reliable. Desert and highland cultural systems are as threatened as low-lying coastal ones. Communities in already marginal areas are those primarily exposed to extra risk.

Half of Mongolia's 2.7 million people depend on livestock – horses, cattle, goats, sheep – and a pastoral way of life over a vast, cold, high, desert region.

Recently, rains in summer are lighter, snows arrive earlier, spring storms are more violent, glaciers retreating and permafrost melting. *Dzud*, a famine induced by snows falling too deep for animals to forage for grass beneath them or by ice coating the soil, used to be a three-times-a-century occurrence. No longer. In spring 2000, 18 per cent of the herds – five million animals – died because of *dzud*. Mongolia also suffered the chaos of the post-Soviet era, with a pull-out of support and the privatization of herd ownership increasing numbers and over-grazing. But here as in many fragile environments, climate change is helping to pin an ancient way of life to the wall.[6]

Meteorological volatility

The impact of climate crisis on all forms of life is mediated through water. The consequences of continuing warming of the earth will be immense to all livelihood systems: none will escape.

A strange aspect of global warming is what seems like meteorological inconsistency. The average temperature has risen, with three of the hottest years ever occurring in the last decade, and the ice-cap and snow cover are decreasing; but rainfall has also risen, and snowfall in the northern latitudes has risen most of all. Erratic weather patterns – for example, increased intensity or day-to-day unpredictability of the seasonal monsoon – disrupt planting, cropping and herding patterns for farmers everywhere.

The changes predicted in water supply, agriculture and the survival and reproduction of species are very wide-ranging and depend on local variations of climate and hydrogeology. Rivers now fed by snowmelt may swell at different times of year if they are fed by rainfall instead. In highland areas of Asia and the Americas, lakes may fill to an unaccustomed brim with melted snow and ice, threatening to send cascades of floodwaters into heavily populated valleys. Studies of river basins in China, Australia, Belgium, Switzerland

and the US find that rain will be more intense and floods more likely. In Africa by contrast, where the interior is mostly sub-humid or arid, run-off in major river basins has declined by 17 per cent in the past decade, lowering soil moisture and suggesting a disastrous cranking up of existing food insecurity.

Throughout the world, aquatic ecosystems will be affected by changes in water temperature, with some species attempting to migrate while others invade and take over. Certain disease agents or vectors may flourish where they did not flourish before, and outbreaks of waterborne or insect-borne infection overcome populations with no previous immunity.

Already, according to Andrew Simms of the UK's New Economics Foundation, the number of environmentally displaced people forced to leave their homes as a result of extreme weather, drought and desertification is higher than those evicted by political insecurity: 25 million compared to 22 million. By 2050, some island and delta states may have vanished, driving the numbers up to 150 million. What happens to those whose nations become uninhabitable? 'Should they have new lands carved out for them? Or should they become the first true world citizens? If there is no state left, how can the state protect its citizens?'[7]

The reduction of greenhouse gases by cutting back on emissions from the burning of fossil fuels is the only way to tackle increasing climate instability. The US is the largest and most intransigent atmospheric polluter, contributing 25 per cent of total emissions. The ostrich-like attitude of the George W Bush administration and its refusal to take part in international regulation relating to climate change is notorious. Political failures in this regard are seen by some scientists as more threatening to humanity than those associated with the war on terror.

Threats to water quality

The profligate consumption of oil and coal is only one of many pollution threats from industrial processes currently facing our freshwater resources. Traditional craft industries use water, but rarely on a grand scale. And the huge volumes of water extracted for modern manufacturing processes have to be expelled again – in an inferior condition.

In 1969, the Cuyahoga River in Cleveland, Ohio, became so contaminated by industrial effluent and chemical pollutants that it caught fire. This scare led in 1972 to a US Clean Water Act, ushering in the era of polluter regulation. No longer did US industries have the right to pour into waterways whatever discharges they wished, destroying plant and fish life and filling wild water with poison. Gradually, efforts to clean up lakes and rivers took off in North America and Europe. But elsewhere – Eastern Europe, the ex-USSR, and increasingly in industrializing Asia – many waterways remain loaded with sulphurous and stinking contaminants as well as raw sewage.

Today's pulp industry uses 60,000 to 190,000 gallons (1 gallon = 3.78 liters) of water per ton of paper or rayon. Bleaching uses 48,000 to 72,000 gallons per ton of cotton.[8] And these amounts are negligible compared to the requirements for the automobile and computer industries. One estimate of the water needed by a state-of-the-art silicon wafer plant in New Mexico is over 4.5 million gallons of water a week.[9] The amount and proportion of water used in manufacturing is rising as industrialization speeds up around the world. So the quantity of industrial effluents discharged into freshwater ecosystems has been rapidly rising too, from around 237 billion tons in the mid 1980s to 468 billion tons by 2000.

Acute and chronic problems

There are many examples of acute crisis. In 1999, a cyanide spill into the Danube in Romania poisoned

The big polluters

Many substances are discharged by industries into the world's rivers, lakes and aquifers, using up vital oxygen in the water. These include heavy metals such as lead, cadmium and mercury, and some of the most dangerous chemicals ever created, persistent organic pollutants (POPs). These are carbon-based, persist in the environment for a long time, concentrate in the food chain, travel long distances and are linked to serious health effects.

Many chemicals are disposed of onto, under or into the ground, and travel through soil into aquifers. In the US, 60% of liquid hazardous waste is injected into the ground, and traces have turned up in aquifers in Florida, Ohio and Texas. Worldwide 300-500 million tons of heavy metals, solvents, toxic sludge and other waste accumulate in water sources.

Polluting industries

Share of organic water pollutants by industrial sector, 1990s

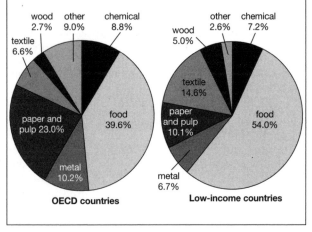

OECD countries

- wood 2.7%
- other 9.0%
- chemical 8.8%
- textile 6.6%
- paper and pulp 23.0%
- food 39.6%
- metal 10.2%

Low-income countries

- other 2.6%
- chemical 7.2%
- wood 5.0%
- textile 14.6%
- paper and pulp 10.1%
- food 54.0%
- metal 6.7%

Atlas of Water, Earthscan, 2004

the drinking supply for thousands of people. In the 1970s, the notorious Love Canal episode in Canada nearly left the entire Niagara and St Lawrence system infiltrated by toxic sludge. There may be slow build-up problems: seepage into groundwater from factory-farm sewage lagoons now common throughout North

America; the high concentrations of nitrogen in modern fertilizers washing into streams; or the steady drip of toxic effluent into the waters and soil of the *maquiladora* or export-processing free trade zones along the US-Mexican border. Here the production of chemicals has tripled since the North Atlantic Free Trade Agreement (NAFTA) was signed in 1994.[10]

In the industrialized world, the principle of 'polluter pays' is upheld, more or less effectively. Some newly industrializing countries give lip service to this idea, but are unwilling to hinder industrial and economic performance or lack the means of regulation and enforcement. Rapid industrial growth in China has stifled pollution control, as it has elsewhere. Today, 80 per cent of the country's 50,000 km of major rivers are so degraded that they no longer support fish. In the Yellow River, discharge from paper mills, tanneries, oil refineries and chemical plants means that the water is now laced with heavy metals making it unfit even for irrigation. Traces of lead, chromium and cadmium have been found in vegetables sold in city markets. Farm chemicals washed into the sea cause massive blooms of algae.

Widespread concern is now routinely expressed for the degradation suffered by inland waters. In the Mekong River, inland fisheries have dropped by half in the past 25 years. In South America, Colombia's Magdalena River has seen its fisheries decline by two-thirds in 15 years because of pollution. Worldwide, of the creatures associated with inland waters, 24 per cent of mammals and 12 per cent of birds are threatened, as are a third of fish species.[11] Industrial contamination is not the only cause: pollution from untreated sewerage is also important, as is the dam-building and canal-building discussed earlier in this chapter. Gradually, it is being borne in on us that technology cannot replicate the many functions our freshwater ecosystems perform more effectively and for nothing, if we let them. Better environmental management of our water bodies, for their and our own conservation, is a must.

The meaningless maze of regulations

In most parts of the industrialized world, there are now extensive testing and regulatory regimes for water in all its environments. In the less industrialized world, such regimes tend to be less developed, and – in common with other regulations without which the industrialized lifestyle cannot be safe – less efficiently applied. In places where democratic institutions are non-existent or only superficial, the power of those affected to demand clean-ups of hazardous waste or proper regulation of industrial or farm pollution is limited. Where a state-run factory in Russia discharged radioactive particles into a river over three decades, or in countless other locations where arsenic, nitrates, chloride, polychlorinated biphenyls (PCBs), mercury or lead have caused sickness and deformity, there is usually no redress.

Recently, there has been a hard-won success against the pollution of commercially-sold drinks and bottled water in India. Here, there is a strong environmentalist movement and many authoritative figures are involved. In February 2003, the Centre for Science and Environment (CSE) published the results of a study into bottled water showing that many brands sold under an Indian Standards quality certification were laden with pesticide residues. CSE followed up with a report that Coca-Cola, Pepsi and leading brand soft drinks were many times over European Community (EC) limits for lindane, DDT, malathion and other carcinogenic and immune-suppressant pesticide components.[12] There was a particular irony about the discovery of toxic ingredients in products marketed specifically on the basis that they were safe to drink.

A row immediately erupted between the soft drinks manufacturers and the CSE, accused of spiking its chemical tests. Coca-Cola and PepsiCo – who account for 80 per cent of an annual $1,650 million cold drinks market in India – were aggrieved at the suggestion that they manufactured products with toxin levels far above the norms permitted in Europe and North America. A Parliamentary committee was set up to look into the matter, and reported in February 2004 that five laboratories carrying out similar tests had confirmed CSE's findings.[13] Their report recommended stringent new regulations.

The CSE had exposed glaring holes in the standards governing their manufacture. 'In India, these companies cannot be taken to court. In fact, forget legal procedure; these companies cannot even be politely told to stick to norms. For – and this is precisely where quality-conscious multinationals laugh all the way to the bank – the norms that regulate the manufacture of cold drinks in India are a meaningless maze,' wrote CSE's Sunita Narain.

The toll of disease in India and around the world from water- and pollution-related causes is still overwhelmingly connected to bacteriological contamination. But there is an important implication in the CSE's findings. The companies drew their supply of raw water from groundwater at their premises. The bottled water tests showed that the pattern of pesticide contamination followed, in weaker doses, the pattern of contamination in the groundwater at these sites. Therefore the groundwater may be similarly contaminated in many parts of the country. Groundwater is now used for drinking by 90 per cent of rural consumers in India and 50 per cent of urban consumers;[14] it has specifically been exploited for this purpose because it is meant to be safe. In the world as a whole, 25 per cent of people rely on groundwater for drinking purposes.

What price safety now? In India, and where similar policies are followed elsewhere, what was once a 'safe' supply of water is now threatening to pose a whole new generation of health and disability problems down the line. The regulation and treatment to prevent this is regarded as 'unaffordable'. The destruction of freshwater ecosystems will mean the destruction both of future livelihoods and of future human health. To regard these losses as 'affordable' seems as obstructionist to planetary health as the official US attitude to the Kyoto accords.

The battle of the bottle

The growth of the bottled water industry when it began to take off in Britain and North America around 20 years ago was partly a reaction to the perceived unhealthiness of the piped supply of water for drinking. As environmentalists began signalling the dangers of chemicals, pesticide residues and heavy metals in our surroundings, not only did we stop swimming in our rivers but we stopped believing that goodness flowed from our taps. Bottled waters began to be sold more actively on the back of claims to the purifying and health-conferring qualities of their natural sources. Although costing several thousand times the price of tap water, people seemed willing to be seduced in droves into this new lifestyle option.

In the 1970s, the annual volume of water bottled and traded around the world was about one billion liters. This more than doubled by the end of the decade, but the real growth began in the mid-1980s when the multinational companies, led by Nestlé, Coca-Cola and PepsiCo began to realize that there would be a huge market in the large parts of the world where the water supply was never bacteriologically safe. By 2000, the amount of water bottled and sold had risen to 84 billion liters. Worldwide, sales in bottled water were estimated at \$22 billion in that year. This is an astounding rate of growth for any industry. When you realize that, not only in a country such as India, but in the US and Britain as well, bottled water when tested may turn out to be less safe than tap

Global bottled water sales

Sales of bottled water have grown dramatically over the past decade. In the US, they have persistently grown by 10% a year since the 1970s. Growth rates are now even higher in some parts of the world, notably in China, Eastern Europe and CIS.

Country/Region	1996 Sales (Million liters)	Projected 2006 Sales (Million liters)	Annual Percentage of Growth
Australasia	500	1,000	11
Africa	500	800	4
CIS	600	1,500	13
Asia	1,000	5,000	12
East Europe	1,200	8,500	14
Middle East	1,500	3,000	3
South America	1,700	4,000	7
Pacific Rim	4,000	37,000	18
Central America	6,000	25,000	11
North America	13,000	25,000	4.5
Western Europe	27,000	33,000	2.5
Total	**57,000**	**143,800**	

Peter Gleick, The World's Water, 2002-2003

water, it seems extraordinary that consumers are so willing to be had.

The purity and healthy lifestyle case for bottled water is usually underpinned by the notion of its 'pure spring' source. In fact, one-quarter of all bottled water is simply tap water, processed and purified to some extra degree, but often subjected to less rigorous testing and regulation than the original tap water itself. In Britain, when Coca-Cola introduced its Dasani brand in March 2004 and it became publicly known that the water originally came from the mains in Sidcup, Kent, the company was ridiculed. If Coca-Cola had not been over-aggressive in trying to force other 'spring' waters off the shelves, 'Eau de Sidcup' might never have hit the news. The company suffered a second public relations disaster within weeks. The entire Dasani stock had to be withdrawn when it was discovered that, during processing, naturally occurring and inoffensive bromide in the Sidcup tap water had been oxidized into bromate, a carcinogen.[15] It remains to be seen after re-launching whether Coca-Cola will persuade UK consumers to pay 95p ($1.70) for half a liter of dubious water whose original cost, unbranded and perfectly safe to drink, was 0.03p.

The bottled water saga is not confined to quality. Like so many other issues to do with this threatened resource, quality and quantity often turn out to be two sides of one coin. The commercial search for 'pure spring' – or just uncontaminated – sources to pump up and bottle exhaustively is having other damaging effects. In Uruguay, foreign-based corporations are buying up wilderness tracts and even whole water systems for future development. Often, the companies extract huge quantities of water while paying little or nothing for it, and may cause environmental mayhem in the bargain.

In another landmark case in India involving Coca-Cola, the company was recently ordered by the high court in the southern state of Kerala to stop drawing on a groundwater aquifer. Villagers complained that its bottling plant was drying out the land, destroying their coconut palms and turning their rice paddies into a dust-bowl. Against the wishes of the state government they took the company to court. In December 2003, Justice K Balakrishnan Nair found in their favor, and in so doing made a very important statement. 'Groundwater under the land of the company does not belong to it. Every landowner can draw a reasonable amount of water necessary for domestic and agricultural requirements. But here, 510,000 liters is extracted per day, converted to products and transported, thus breaking the natural water cycle.'

His statement was music to many environmentalist ears. There is a genuine fear in many parts of the world including India that rights in water are being eroded by concepts such as 'management by demand' and 'willingness to pay', and by the might of the corporations who know how to twist arms and legal regimes to gain unrestricted access to the resource from which they make money. Who owns the contents of lake, stream or aquifer? Who owns the rain? The defence of water as a commons over which all humanity has rights is the next subject we'll address.

1 Diane Raines Ward, *Water Wars: Drought, Flood, Folly, and the Politics of Thirst*, Riverhead Books, New York, 2002. **2** Sandra Postel and Brian Richter, *Rivers for Life: Managing Water for People and Nature*, Island Press, Washington, 2003. **3** Sharier Khan, 'Bangladesh Blames India for Destruction of Sundarbans', *OneWorld South Asia*, 23 March 2004. **4** *Natural Catastrophes, Infrastructure and Poverty in Developing Countries*, International Institutes for Applied Systems Analysis, 1999; quoted in Constance Elizabeth Hunt, *Thirsty Planet: Strategies for Sustainable Water Management*, Zed Books, London, 2004. **5** Data from the Center for Research on the Epidemiology of Disasters (CRED), Brussels, quoted in *Water for People, Water for Life*, UN World Water Development Report, UNESCO-WWAP, 2003. **6** Lutaa Badamkhand, 'Mongolia: Winter Disaster Kills Herds, Livelihoods', IPS, 24 April 2000. **7** Andrew Simms, 'Unnatural disasters', *The Guardian*, 15 October 2003. **8** *Climate Change 2001*, International Panel on Climate Change, Cambridge University Press. **9** *Sacred Waters* (1997), South West Network for Environmental and Economic Justice, quoted in Vandana Shiva, *Water Wars*, Pluto Press, London, 2002. **10** Maude Barlow, Tony Clarke, *Blue Gold: The battle against corporate theft of the world's water*, Earthscan, London 2002. **11** *Water for People, Water for Life*, UN World Water Development Report, UNESCO-WWAP, 2003. **12** *Down to Earth* magazine, Vol 11, No 18, 15 February 2003 and Vol 12, No 6, 4 August 2003, CSE, Delhi. **13** Randeep Ramesh, 'Soft-drink giants accused over pesticides', *The Guardian*, 5 February 2004. **14** Ashok Nigam, Biksham Gujja, Jayanta Bandyopadhyay, Rupert Talbot, *Freshwater for India's Children and Nature*, UNICEF and WWF, New Delhi, 1998. **15** Felicity Lawrence, 'Things get worse with Coke', *The Guardian*, 24 March 2004.

6 Rights and wrongs of water

Water rights are traditionally rights of access and of use. Legal systems throughout history have recognized that water is special and cannot be privately owned. But today, the idea of water as a natural commons to be managed for the public good is under threat. In many parts of the world, the state has failed in its duties as the custodian and regulator of freshwater resources. The entry of commercial entrepreneurs as surrogates for the task erodes water rights still further.

SINCE ANTIQUITY THERE have been rules about water. Questions about who had rights over the stream, the well, the pump or the water wheel were important social matters. As settlements grew, those who interfered with flowing water by building weirs to trap fish, by diverting part of the river away, or by discarding the contents of their cess-pits into it, could not be allowed to do so with impunity. Rulers who built dams, tanks and hydraulic works, needed regulations to govern their management. The Romans, who built aqueducts all over southern Europe, had well-developed water laws.

The codification of rights and responsibilities around water could not be approached in the same way as land. Land is fixed so it can be demarcated by hedges and walls. It therefore became the foundation of private and state property, wealth and inheritance. By contrast, water had to be a communal asset because it did not stay still. Even in a lake or pond, water is on the move – no part of it can be permanently marked out. For thousands of years, legal systems across the world accepted that there could be no ownership of running water.[1] Ruling authorities had important roles as water's custodians, legislators of rights to its access and regulators of its deployment. Many legal systems acknowledged special rights of owners of land above or alongside water. But the water itself was regarded as a genuinely common asset.

Over the past few years, as the sense of water crisis has become more palpable, the level of competition over resources more acute and the role of the private sector in water provision more pervasive, the sacrosanct idea of water as a commons has come under threat. According to Vandana Shiva: 'The globalizing economy is shifting the definition of water from common property to private good, to be extracted and traded freely.'[2] Hers is far from being the only voice to be raised against that part of the neoliberal agenda which proposes to substitute pricing mechanisms and the creation of markets for publicly-run protection and regulation of our common freshwater resource.

The idea of water as a natural commons is ecologically sound and socially attractive. But in today's exceedingly unnatural world it is difficult to hold onto from a practical point of view. The use of raw, untouched and unprocessed water is increasingly rare and – in the context of public health – regarded as undesirable. Many things happen to most water today before it is drunk, bathed in, pumped onto the fields, sucked into an industrial plant or spewed out of it, and those things require moving this heavy material about and passing it through a plant of some kind. Thus, holding onto the distinction between the resource itself and the contraptions surrounding it has become very difficult. After all, these in their many incarnations are what make the resource accessible and usable to most of its customers. The tendency is to value them ever more highly, in the name of disease control, power generation, irrigation, navigation, flood management, and – ironically – the restoration of water to its pristine or natural quality.

Nature did provide only the water and not the pipes. But those who own the pipes and the many other operational works and run the institutional structures automatically assume control over the water too. When ownership of the works and structures shifts from the public domain to a private commercial enterprise, it becomes much harder to assert people's common ownership of running water: to all intents and purposes the water in the pipes is owned by the enterprise and becomes its means of profit. How can people's rights to access and use of water then be preserved?

Water regimes in history

When the Emperor Justinian tidied up Roman Law in the early 6th century, his statutes declared that running water, like the air and the sea, was held in common and could not be owned. But there was some classification of public and private rivers. Larger rivers were public and had to be open to navigation; small streams could be privately held. At this stage and for a considerable time to come, rivers – and, later, canals – were key communications links between communities and vital arteries of transport; in some places they still are. Many early legal provisions about common rights in water had to do with navigation.

Another key issue of early water rights was to do with food and livelihood: fishing. In Britain, laws of 700 years ago established closed seasons for salmon fishing in both Scotland and England, showing that the river's need to renew its aquatic ecosystem was already understood. Magna Carta (1215), a landmark agreement between the English king and his subjects, included a provision that all permanent fish weirs should be removed from the Thames to allow boats and barges to move freely.

Other matters soon intruded. The medieval history of Foxton, a Cambridgeshire village, shows residents routinely fined for widening or diverting the brook, their way of trying to take more than their share of water from the common stream. Those in the upstream end of the village had to be stopped from fouling the water with their livestock or otherwise impairing its downstream flow for others. The legal principle to emerge was that 'the polluter pays', a principle which paved the way for the modern idea of monetizing emissions and setting up markets in pollution allocations as a method of keeping outputs of toxic material under control.

Early water regulations from many societies recognize 'riparian rights': the rights of people living along river banks to use water in their local streams. But they were also bound to help keep channels and drains in good order and lend their labor for hydraulic works. The *Arthasastra*, a treatise on government by Kautilya, the right-hand man of India's first Emperor Chandragupta Maurya (321-297 BC), laid down all sorts of obligations and stipulations about the construction and management of water works for irrigation.[3] The Indian subcontinent abounds in examples of traditional water management systems, which required not only technical understanding but administrative, tax-levying and local adjudication systems. They confirm the age-old acceptance of natural rights in the resource both on the surface and below ground, and equally of obligations towards managing it in the community's interests.

Water democracies in action

A striking example of democratic local water governance comes from Bali, where sophisticated village irrigation systems have existed for centuries. All farmers who take water from the same stream or river belong to an organization called a *sebak* which meets every 35 days and has its own system of law. The *sebak* plans planting times, distributes water between the members equitably, and fines those who cheat.[4]

A similar 'water parliament' operates in the Arvari River basin in Alwar, Rajasthan. The 72 villages in the basin have re-greened their land in the past decade by building thousands of *johads* – small earthen dams – along the valley bottom. Their parliament for water affairs meets every month and agrees rules about water extractions, what crops may be grown, and so on. Fines are imposed: 20 rupees (US 40 cents) for cutting down a tree, 50 for not reporting that a tree was cut down.[5] This case of local water management in action illustrates that regulatory and managerial arrangements are as important as technical prowess for just and efficient water management. Large-scale engineering projects have tended to obscure this by abrogating power to central bodies and to their technical and contractor satraps.

Examples of water democracy are not confined to the South. The same type of local water management body has existed for over 1,000 years in Aragon, Spain. Water here is regarded as belonging to all farmers, large or small, whose land it serves and each belongs to a water users' association or *comunidad de regantes*. A structure of representative bodies culminates in a *Tribunal de las Aguas* (Water Tribunal) for a zone. The most famous *Tribunal de las Aguas* still survives in Valencia. It meets every week on the Cathedral steps, discusses infringements, sorts out disputes and apportions water in times of scarcity. The tribunal is said to have existed continuously since it was established by the Moors in 960 AD.[6]

Joachin Costa, a 19th century agrarian reformer, described the water *communidades* of rural Aragon as: '... examples of solidarity and social co-operation of a truly marvellous delicacy and perfection, unequalled by the most complex works of precision engineering.' This theme of intricate systems of defining and adjudicating water rights among users of a common resource is echoed from many parts of the world, including the Andes, Mexico, South-East Asia and Africa. It is a riposte to the idea that common ownership of a resource over which there is strong competition is impracticable, and that only private ownership can lead to sound resource management and peaceful co-existence. This neo-conservative idea has exerted a powerful hold over those who regard statements about 'the commons' as romantic delusions or ideological anathema.

Do such social and political arrangements stand any chance of surviving? Within the range of a common catchment or river valley, the idea of a commons seems to function effectively where the necessary mechanism is in place, and where there is no dispute between it and other local administrative and political entities. That, of course, can be a tall order as vituperative stand-offs between groups sharing rivers or straddling aquifers – Israelis and Palestinians, for example – frequently show. Many environmental campaigners believe that it is essential to talk up the principle of community management of a common resource more widely, and institute linked systems of decentralized 'water democracies'. Only then will conflict be defused, and water resources be sustainably and fairly managed.

Water rights undermined

Community rights and responsibilities around the management of streams and wells began to be eroded historically as the state assumed more control over water resources – and over the rest of the natural resource base. A particular case was the American West, where the 19th century authorities – barely distinguishable from the cowboys and buccaneers among whom they represented the law – connived with private entrepreneurs to establish rights over water.

Here, the traditional riparian principle – that people living on the banks of a river had a right to use its flow but only so far as they did not damage it for others – was not applied. Instead, a doctrine of 'prior appropriation' emerged, which meant that those who arrived first and put their mark on something – land, gold, water – could claim permanent rights over it. This made possible the trade in water rights and the development of water markets which is now held up by the corporate lobby as the route to water conservation and pollution control. However, it left the fate of many smaller farmers hostage to fortune. The saying goes: 'In Colorado, water flows uphill towards money'.

Even where the riparian doctrine prevailed, the growth of government – whether colonial or domestic – and its increased involvement in the ordering of all kinds of affairs meant that local systems of water resources management were displaced. And once major engineering works and centralized finance and planning came into the picture, the provision of water services and irrigation works were co-opted into the public realm. Indeed, they were emblematic of the state's advance and accumulation of power – whether for good or ill. The Bengali poet Rabindranath Tagore wrote a play in the 1930s called *The Liberated River* in which he symbolized colonial rule through the dam, and Gandhi's struggle for freedom as the liberation of the river.

Many Southern environmentalists who criticize the centralizing tendencies of colonial rulers as having brought ruin on community-managed systems admit that their modern successors fell even more deeply in love with central planning and grandiose engineering. The newly independent state had become the repository of law, administration and responsibility for the commons – and of patronage and lucrative contracts for construction. Typically, the state initiated water schemes and offered the community free services. This, in retrospect, was 'water supply welfare' and it happened all over the post-colonial world.

Unfortunately, it not only promoted inefficiency, waste and water-profligate land use, but also created dependency. Local systems for irrigation and water supply tended to collapse and existing rights to erode. Schemes not grounded in community needs and local management invariably failed; but this lesson took time to absorb and by the time it did the old ways and the old rules were difficult to recover.

Water management increasingly centralized

The centralization of water management was powerfully enhanced by the post-colonial orgy of dam-building. In 1949, only around 5,000 large dams existed worldwide; by the 1970s construction had accelerated to such a pace that 5,000 were built in a decade.[7] Exercising their custodial muscle, authorities in Asia, Africa and Latin America commandeered plains, valleys and riverine environments, displaced their inhabitants and disrupted their ecosystems, in the name of 'the greater common good'.

The speciousness of this claim was increasingly exposed by the rise in anti-dam protests, which began to capture international attention during the 1980s. In the Philippines, an uprising of indigenous people took place in 1981 against a World Bank-funded project to build a dam on the Chico river. This ignited a worldwide crusade to save similarly threatened communities and brought into being a California-based campaigning and information-generating organization, the International Rivers Network. The many popular protests against large dams which have since gained strength from solidarity and international action are essentially political movements by groups whose resource base is being stolen and who are often too marginal to have any voice in mainstream political life.

The largest, best-known and longest-running anti-dam people's movement is the *Narmada Bachao Andolan* (NBA, literally 'save Narmada campaign') in India's Narmada Valley. Led by Medha Patkar, the NBA began opposing the social devastation to be caused by the construction of large dams on the Narmada River in 1985. Their principal campaign has been against the massive Sardar Sarovar dam and its 75,000 kilometer canal system, touted by state politicians as the lifeline of drought-prone Gujarat. The movement spectacularly achieved the withdrawal of World Bank funding from the Sardar Sarovar in 1993, and has won delays in construction and better

Be dammed

Large dams are defined as over 15 meters in height or with reservoirs containing 3 million cubic meters of water. However, those of the monster variety are well over 100 meters tall and have reservoirs containing thousands of millions of cubic meters. The reservoir of the Aswan High Dam, for example, contains 168,000 million m^3. Such projects, where they submerge the lands of hundreds of thousands of people and cause vast ecological change, are only practicable in settings where power is heavily centralized and can ride roughshod over the interests of local populations.

Regional distribution of large dams

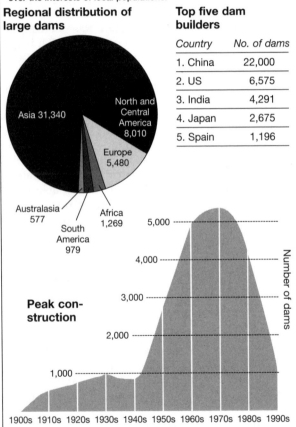

Asia 31,340

North and Central America 8,010

Europe 5,480

Australasia 577

South America 979

Africa 1,269

Top five dam builders

Country	No. of dams
1. China	22,000
2. US	6,575
3. India	4,291
4. Japan	2,675
5. Spain	1,196

Peak construction

Number of dams

5,000
4,000
3,000
2,000
1,000

1900s 1910s 1920s 1930s 1940s 1950s 1960s 1970s 1980s 1990s

World Commission on Dams, 2000

deals for the displaced. Even if it has not managed to stop the project altogether, it has been a huge inspiration to similar people's movements. These campaigns and their attendant publicity succeeded over time in bringing the whole dam-building and river-dominating ethos into disrepute.

Gradually it became inescapable that the state was failing as guardian and regulator as well as service provider and manager. Its poor performance of its custodial role helped open the door to the corporate invasion. Certainly, reform was needed. But devolving some of its functions onto private commercial operators required that the state become stronger as a regulator and defender of water rights.

Unfortunately, the opposite has tended to happen: a weak, inefficient and sometimes venal bureaucracy hands over its resource management role to its cronies and looks the other way. Slimmed down by retrenchment in the pursuit of 'smaller' government which is another reform and structural adjustment rule, its capacity is anyway depleted. In many settings, the chances that the regulatory framework will be used to penalize polluters, raise the tariffs of water-guzzlers, and champion ordinary users' rights against commercial and industrial proprietors are virtually non-existent.

Water management increasingly monetized

Once the resource itself has effectively become a tradable commodity it is a short step to allowing its governance to be dictated not by the doctrine of common good, but by the regime of private property rights underpinning the buying and selling of goods. Market enthusiasts point to the Murray-Darling basin in Australia, where a system of water trades and credits, carefully regulated, has revolutionized water use patterns of towns and farmers. But what works in Australia cannot be used as a template for equally arid parts of Africa or Asia: everything about the

Australia's great experiment

The Murray-Darling basin covers a territory as large as France and Spain combined, and contains 20 major river systems and a labyrinth of wetlands. Half of Australia's agricultural output is produced in the basin, also home to half the country's sheep and a quarter of its cattle. But decades of irrigated farming and herds on the hoof ruined the fragile soils and led to serious water-logging and salinity.

In 1987, South Australia sued upstream New South Wales to stop it issuing more water allocation licenses and it was agreed that the four states served by the basin had to take concerted action. Decision-making powers were conferred on the Murray-Darling Basin Commission to manage land, water and environmental resources in the entire region.

The Commission put a number of programs into action – not a moment too soon. Loss of production was estimated at $390 million a year, and pollution from farms, factories and sewage on the upper Murray and Darling was out of control. In 1992, this produced the largest bloom of blue-green algae ever seen, killing fish and poisoning livestock along the Darling's banks for 600 miles.

Salinity and water-logging have since been curbed by limiting rice production, insistence on deep drainage and water-saving sprinkler or drip irrigation, tree-planting, and a system of salt credits whereby those who emit less salt than their agreed quota can trade it to others. The profits are ploughed into salt removal: 1200 tons are trucked out of the Murray every day.

At the heart of the Commission's success are the big money stakes which act as a powerful motivation, and determination to structure the body so that not only key state stakeholders but the population of the whole area can participate. Each river system has its own catchment committee, supported by community groups. The Commission framework allows every problem to be addressed by a cross-section of parties. It has managed to put a cap on extractions. No-one withdraws water or creates pollution without approval or restitution.

Although such a model cannot simply be transferred elsewhere, it is reassuring that democratic and environmentally-conscious water management is practicable over such a vast river basin. And it shows that, where the political, regulatory and economic environment is suited, market mechanisms can play some role in water management. It does not help their opponents' case to deride them absolutely.

Diane Raines Ward, *Water Wars*, Riverhead Books, New York, 2002 socio-economic and politico-administrative context is different.

In very different environments – and they don't have to be only in the South – the tendency will be for water to flow towards whoever has money. Examples are legion from all over the world: from France, where IBM taps groundwater aquifers for IT manufacture which ought to be state-protected;[8] to Ghana, where the poorest communities contribute capital and maintenance costs for water services they cannot afford to buy, and the state cannot afford to administer. In societies with vast socio-economic discrepancies and frail democracies, the disadvantaged may be drained dry – by a soft drinks company for example (see Chapter 5) – or suffer water exclusion.

The point is best illustrated by an Indian example. In the state of Chhattisgargh, the rights over a 23 kilometer stretch of the Sheonath River have been leased out to a private entrepreneur under a build-own-operate-transfer (BOOT) agreement.[9] This is legally a 'concession', but the businessman-operator, Kailash Soni, has a monopoly control over all the water in the river via the structure he has built – called a Flood Regulating Barrier System – and all the water passing through this structure is effectively his. Distilleries and food businesses have been attracted to the local industrial park by Soni's reliable and cheap water supplies. But the villagers who live along the banks of the river and have for centuries fished in it, bathed in it and used its waters to irrigate their crops, are now only allowed access to it at Soni's whim. His entrepreneurial success has inspired a host of similar projects across the country. But it has also inspired accusations of rights violations on behalf of the villagers.

If people who, for centuries, have used the waters of a given river to support their lives and livelihoods do not 'own' it, then how can a company with a license from the state exclude them and behave as if it does? In the face of such developments, it is safe to assume that water safeguards for those who cannot easily defend themselves through the political process will be sacrificed. No wonder a new regime of rights around water is required.

The human right to water

During the 1990s, the defence of human rights assumed a central role in the development discourse. Instead of being confined to the civil and political rights protests of the Cold War era, campaigns now championed social and economic rights as a powerful rationale for equitable development policies. Social activists began trawling through international human rights legislation to see whether there was some way of putting pressure on the state to supply this service or that by advocating fulfilment of human rights.

With consternation it was discovered that no human right to water had ever been articulated. Only in the most recent human rights treaty, the 1989 International Convention on the Rights of the Child, was the right to drinking water and sanitation asserted. The rights-enthused therefore claimed that it could be inferred from other rights articulated in the Universal Declaration of Human Rights (1948) and the International Covenant on Economic, Social and Cultural Rights (1966): the rights to life, to food, and to a standard of living adequate for health and well-being. Promoters of 'water as a human right' argued that water, like air, was so fundamental a resource that the drafters of these older conventions had not thought it necessary to mention it explicitly.[10] However, what they were referring to was drinking water or 'welfare water'; not to rivers, aquifers, or water in a livelihood context.

In 1996, the new South African constitution gave a boost to the claim that it was necessary to recognize the human right to water. Some other African countries undergoing political upheaval and constitutional revision followed suit. The new South African water law, passed in 1998, states that the right to water means that minimum human needs, together with a basic environmental reserve, have first claim on the country's water resources. This 'basic minimum' concept has since been echoed in other water laws and policy statements.

South Africa also set out to extend water services to citizens ignored during the apartheid era. However, in order to recover costs and promote conservation by pricing, the government invited in Suez and Bi-Water to privatize services. Water rates rose and many could not pay. In Johannesburg, citizens in the townships resisted cut-offs by pulling out their meters and hooking themselves up illegally. In KwaZulu Natal, during 2000-2001, 100,000 people became ill with cholera and 220 died as a result of cut-offs.[11] It is one thing to enshrine rights in constitutions and another to deliver them on the ground. The first 25 liters per day of water to a household are now provided free.

One problem is to do with what constitutes the right to water in terms of amount. The amount of 25 liters for a household is very small: around five liters a head. This may be enough not to die of thirst, but precious little else. Others have proposed a daily minimum of 50 liters per head, but where a family is growing vegetables, brewing beer, nurturing livestock, making cloth or food for sale, this will not go very far. Despite the difficulties of definition, the campaign for a 'right to water' has had some success. In November 2002, the UN Committee on Economic, Social and Cultural Rights made good the lack of an explicit right to water in international legislation. They put an obligation on governments to 'progressively extend access to sufficient, affordable, accessible and safe water supplies and to safe sanitation services'.[12]

The articulation of the human right to water is a step forward, even though its application will be fickle and is confined to drinking water and minimal domestic needs. NGOs welcome the opportunities it provides for advocacy on behalf of the 1.1 billion people still without a safe supply of drinking water, even though by itself it does not necessarily bring their access closer. More problematically, it does nothing for those smallholders and indigenous peoples who are deprived of their natural resource base by dam and canal construction or whose environment is ruined by water over-extraction. In the end, the defence of their rights depends more on legal regimes governing land and water running under and through it than those set up to defend humanity.

However, any gain in the articulation of rights is a potential new line of defence for the downtrodden and dispossessed. Some countries, under pressure from their environmentalist lobbies, are taking the right to water seriously. One example is Argentina, where cases have actually been brought to court to defend the rights of local indigenous people against the provision of a water supply contaminated with lead and mercury.[13]

Sharing rivers

The most difficult political issues around water emerge from the clash of upstream and downstream users' interests. Competition between urban users and farmers has led to great bitterness in the American West, notoriously when Los Angeles managed to steal the contents of Owens Lake – the story of Roman Polanski's movie *Chinatown*.[14] Water in California, Arizona and Nevada remains a murderous subject. Arguments around allocations from the Colorado River to these three of the seven states in the 632,000 square kilometer basin remain fraught. No number of agreements – several were brokered in 1998 by Bruce Babbit, President Clinton's Interior Secretary and self-appointed 'River Master' – can ever deal definitively with the problem that there is not enough water to provide for an infinite number of swimming-pools and fountains in expanding luxury cities such as Las Vegas. Farmers have begun to trade their water rights to the towns as water itself becomes a more lucrative crop than anything it could grow.

The contest between irrigation and thirsty cities can only grow more acute as the world progressively urbanizes. In northern China, reservoirs that used to supply water to farms are now used almost exclusively to supply households and factories. In Daxing County south of Beijing where supplies flow in from city reservoirs, farmers were already reporting in 1993 that they had not managed to flood a rice paddy for more than a decade. As China industrializes at a furious pace, there is more pressure to pull supplies away from agriculture to reduce urban deficits. Rice farmers in Java face a similar problem, losing their supplies to textile factories even though Indonesian law gives agriculture a higher water priority. Some factories rent rice fields for their irrigation water; meanwhile their discharges cause pollution, lowering crop yields and killing fish. Farmers suffering from lost production end up selling their land and migrating to town.[15]

Water scarcity also pits farmers against other farmers: the dispute between India's Karnataka and Tamil Nadu states over Kaveri waters regularly leads to violence as noted in Chapter 3. Farmers may also be pitted against indigenous peoples. In Klamath Falls, Oregon, during the drought year of 2001 farmers repeatedly took the law into their own hands and opened reservoir gates for irrigation that the FBI had ordered closed. Local native tribes had treaty rights to the fish populations upstream, and demanded that sufficient water be retained to support the natural habitat. The local law enforcement agents sided with the 1,400 local farmers and refused to prosecute trespassers. In a similar dispute in California, the federal court determined that redirecting water to help endangered species constituted 'a taking of property' and ordered compensation for farmers.

Here is yet another issue of rights: how does the

environment champion its cause? Can the birds speak or the trees complain? Can the fish revolt or the mangroves muster? Some states have passed laws to protect endangered species which require river flows to be kept above a certain level. But even where it is legislated, the defence of nature's right to water – without which the continuity of the hydrological cycle cannot be guaranteed – often defaults to environmental activists and indigenous peoples. Their tally of victories is miserable compared to their tally of defeats. Small farmers trying to hold onto their living are rarely the real villains of the piece. Smallholders everywhere – in Wales as in Mexico, India as in Arizona, Kenya as in Brazil – are having almost as difficult a time as indigenous people and pastoralists defending their precarious livelihoods in a commoditized and corporately driven world.

Only in societies with established democracies and sound, clean – 'safe'? – administrations is there much prospect of effective political mobilization in favor of water rights. The arbiters of land use, industry, agriculture, energy, transport and tourism, pursue their water demand and water conservation policies with the blessing of the world's international financial and trading institutions. Unless they can manage it with little extra expense, they are not over-committed to environmental integrity or to sustaining the livelihoods of around 1 billion rural people at the edge. First they invested in water wastefulness; now, having created scarcity, they use pricing and markets to drive small people to the wall, thereby gaining control over a larger slice of the natural resource pie while reaping a convenient harvest in water profits. That may not be the intention, but it is the effect.

Across the boundaries

Worldwide, 269 river systems cross national boundaries. The competing rights of water users within the same political or administrative entity are difficult enough to resolve. But they become infinitely more complex when the upstream users belong to a different state (within a federal structure), or a different country, from the downstream. If these states or nations – and there may be several in one river basin – are politically at odds, the problems associated with their shared water resource can have far-reaching effects on political, social and economic life.

Wars or threats of wars between riparian states, and the implicit possibility of cutting off headwater flows, have had the capacity to concentrate diplomatic minds in a wonderful manner. Take the Indus River, whose downstream flows were divided between India and Pakistan by the 1960 Indus Waters Treaty. The two countries have gone to war on two occasions since then, but on neither did either side stop honoring its obligations on Indus flows and releases. On the other side of the subcontinent Bangladesh's rancor over India's diversion of Ganges water via the Farraka barrage above Kolkata has been bitter. Since the 1970s, sabers were persistently rattled and terrorist raids threatened as Bangladesh's delta-dependent agricultural economy suffered persistent freshwater deprivation. In 1997, the Ganges Water Sharing Treaty finally set down each state's rights and allocations. But this dispute is one that can easily re-open as population and economic pressures build.

The long list of international river basin disputes is headed by the mega-wrangles over the waters of the great rivers which made the Middle East the cradle of civilization: the Nile, the Jordan, the Tigris and the Euphrates. The Tigris and Euphrates both rise in eastern Anatolia in Turkey, and the Turkish attitude is that they have as much right to their water as Iraq does to its oil. Turkey's claim to sovereignty over the headwaters is balanced by Iraq's claim that its historical rights to water go back six millennia to Mesopotamian times, a long-established doctrine of 'prior use'.

Syria, positioned in between, uses the 'water sovereignty' argument to downstream Iraq, and the 'prior use' argument to upstream Turkey. Turkey has $32 billion worth of dams, canals, barrages, weirs and underground aqueducts on the drawing board, intended to green its south-eastern provinces. Delays and costs have postponed completion until 2010. If this occurs, Syria will lose up to 40 per cent of its current Euphrates take-off, and Iraq, 80 per cent. How they will manage is yet to be seen. Meanwhile water disputes in the Middle East are not confined to surface streams: Israelis and Palestinians are in a perpetual state of angry negotiation over the aquifers under the West Bank, as well as wider-scale disagreement over the waters of the Jordan River.

Somehow, transboundary experts remain sanguine about the prospects of countries actually declaring war over water. Despite all the threats and fist-shaking, the fundamental requirement of all life forms for water exercises a restraining effect. One example is the ten African countries sharing the River Nile. For 12 years they have been meeting under the Nile Basin Initiative in an attempt to resolve differences amicably. At their 2004 meeting, the members agreed to form a permanent institutional and legal framework to facilitate basin-wide co-operation in 'integrated water resources planning and management', the current mantra of the international water establishment.

This concept suggests that if water, land and other components of the ecological system are managed as a whole within the watershed or river basin, it ought to be possible to maximize economic and social well-being, resolve disputes between users in an equitable manner, and make sure that ecosystem integrity is preserved. The approach was devised as an antidote to current, fragmented, regimes whereby water for

Water tensions in the West Bank

The 1967 Six-Day War was, in part, Israel's response to Jordan's proposal to divert the Jordan River for its own use. In seizing the area known as the West Bank (of the Jordan), Israel not only gained access to the headwaters of the river, but control of the aquifer underneath, thus increasing its water resources by nearly half.

In the West Bank, many Palestinians are obliged to get by with 35 liters a day, while Israeli settlers nearby enjoy their swimming pool and garden lawn lifestyle. Israel extracts up to 75 per cent of the flow of the Upper Jordan River via its National Water Carrier (a canal and pipeline system), leaving a trickle to reach the West Bank. It has also categorically stated that it is not prepared to share the waters of the Mountain Aquifer (source of three aquifers in the West Bank, the northern, eastern and western) equally with Palestine. There have been times when Israel has threatened to ban all Palestinian well-drilling in the area. With the construction of the Wall, the access of some farmers to their usual wells and pumped supplies will be further constricted.

Since 1993, a common Water Development Program has been in place, and although negotiations are hampered by Israel's refusal to share water equally, officials from both sides continue to meet.

Water imbalance

Average domestic allowance per person per day, 2002

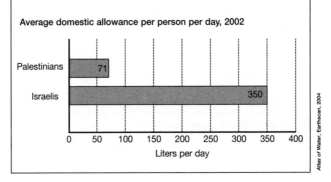

domestic services, agriculture and fishing, water for nature, water for industry, mining, power, navigation and recreation are separately administered and not brought into a common policy framework.

In theory, this idea of holistic management of water

and the ideal of fairness it contains ought to be used to defend every person's fundamental right to water, meaning access to an affordable supply adequate for a healthy and productive life within easy reach. But some of those promoting this agenda are card-carrying members of the neoliberal nexus who are helping to commoditize the resource and promote its governance by market propositions. Can the concept be advanced without being swamped by internal contradictions or hijacked by a water-profiteering cartel?

1 David Kinnersley, *Troubled Water: Rivers, Politics and Pollution*, Hilary Shipman, London, 1988. **2** Vandana Shiva, *Water Wars: Privatization, Pollution and Profit*, Pluto Press, London, 2002. **3** Anil Agarwal and Sunita Narain (Eds.), *Dying Wisdom: Rise, fall and potential of India's traditional water harvesting systems*, CSE, New Delhi, 1997. **4** Robin Clarke, *Water: The International Crisis*, Earthscan, London, 1991. **5** Maggie Black, *A Matter of Life and Health:* OUP and UNICEF, New Delhi, forthcoming (2004). **6** Colin Ward, *Reflected in Water: A Crisis of Social Responsibility*, Cassell, London, 1997. **7** *Dams and Development*, the Report of the World Commission on Dams, Earthscan, London, 2000. **8** Riccardo Petrella, *The Water Manifesto: Arguments for a World Water Contract*, Zed Books, London, 2001. **9** Bhavdeep Kang, 'This Man Owns the River', 16 September 2002, www.outlookindia.com/ **10** Peter Gleick, *The Human Right to Water,* article for *Water Policy*, Pacific Institute for Studies in Development, Environment and Security, March 1999. **11** Maude Barlow and Tony Clarke, *Blue Gold: The Battle against Corporate Theft of the World's Water*, Earthscan, London, 2002. **12** *A right to water: a step in the right direction*, WaterAid, www.wateraid.org.uk/ **13** Juan Migual Picolotti, *The right to water in Argentina*, CEDHA and Rights and Humanity, November 2003, www.righttowater.org.uk **14** Marq de Villiers, *Water Wars: Is the World's Water Running Out?* Weidenfeld & Nicolson, London, 1999. **15** Sandra Postel, *Pillar of Sand: Can the Irrigation Miracle Last?* WW Norton and Worldwatch, New York, 1999.

7 Water wars or water peace?

In the recent past there has been a great deal of fuss about the world's freshwater crisis. How has the international community, both in its formal and its activist incarnations, responded? Are there signs that a new world water order could emerge?

WHEN IN THE late 1990s, the global freshwater crisis was described as leading the world to the threshold of 'water wars', the idea captured a great deal of attention. In international circles, where water diplomacy was growing in stature, the issue rose dramatically on the agenda. Was this a passing excitement or have the issues of water stress and water scarcity moved significantly forward? If so, in what direction?

There was, especially to begin with, a lot of scurrying around at international level. 'World' bodies were set up to debate policy and solve disputes. The World Water Commission and the World Dams Commission both reported in 2000. The World Water Council, the Global Water Partnership and the UN's Collaborative Council on Water Supply and Sanitation beefed up their activities. In 1997 a UN Convention on the Non-Navigable Uses of International Watercourses was agreed. And there were new efforts to form international and interstate river basin authorities and defuse potential confrontations over large-scale water allocations and diversions.

In the past five to ten years, no issue has been more heavily subjected to international conferences, summits, initiatives and campaigns. There were three World Water Forums, in 1998 (Marrakech), 2000 (The Hague) and 2003 (Kyoto); an International Conference on Freshwater in Bonn in 2001; and at the 2002 World Summit on Sustainable Development (WSSD) – the second Earth Summit – in Johannesburg, water supplies and sanitation dominated the agenda. There was even the first World Toilet Summit in Seoul in 2002, preceded by the first international conference on Ecological Sanitation in China in 2001. In 2003 came the International Year of Freshwater and the publication of the first World Water Development Report, to which 23 UN agencies and secretariats contributed.

At a more popular level, the 'water wars' idea shook people out of complacency. A spate of new books appeared on the theme – tagged to the alarmist prospect of a world running on empty and people at each other's throats over the last drops in the tank, exploring the threats to cultures and livelihoods posed by water's commodification.

These were not 'water wars' in the heavy artillery mode, but low-grade, low-intensity conflicts between town and countryside, between large-scale commercialized and family-based agriculture, between water for industry and water for nature, polluters and conservationists. Most of all they were between water lords and water serfs: state or private appropriators of water resources, and common everyday users – fishing people, farmers, people who fetch their water in a bucket, people who pay considerably more than they used to when they turn on the tap, people forced to deepen their wells or pump water from an ever sinking water table at ever greater energy consumption and expense. In a host of different ways and contexts, water had become hot.

Before long, the scarier 'water wars' scenarios began to fade. The international view currently holding the ring is that water resources have rarely, if ever, been the sole source of violent conflict or war.[1] This is not quite the same as suggesting that humanity stops at the brink of draining enemies dry, but that is also the implication: some people even suggest that the need to share waters can be a harmonizing force. Just one major terrorist strike at a large water-retaining structure or the deliberate poisoning of a major water supply would change perceptions about water's role in conflict. But until or unless that happens, water has been downgraded as a *causus belli*, even while tensions around it continue to flare. Much depends, year to year, on the vagaries of the various monsoons and tropical rains, to whose behavior a very large proportion of the human race is regularly in thrall.

Water activism gains momentum

While the 'water wars' idea has temporarily peaked, activism around water from the grassroots to the international campaigning level has grown. One context is the diffuse and multi-faceted anti-globalization movement whose members target the neoliberal agenda promoted by the World Trade Organization (WTO), the International Monetary Fund (IMF) and the World Bank. Their drive to privatize water and sanitation services so as to facilitate their takeover by transnational corporations on concessionary terms has been widely interpreted as an assault by private capital on a resource which ought definitively to remain in the public domain. For this reason, Bechtel's expulsion from Cochabamba, Bolivia, in 2000 was heralded as an iconic anti-globalization victory.

The second key arena for water activism is the ongoing attempt to prevent the construction of large dams and highlight the damage they inflict on ecosystems and human habitats. Rearguard action to protect natural resources – rivers, forests, wetlands, wildlife – against the behemoth of 'progress' have taken on a new intensity. Dam protests continue and at some sites – notably the Ilisu dam in Turkey in 2001 – the clamor has led to the withdrawal of international funding. A by-product of the election of a leftist government in Spain in 2004 was the cancellation of the National Hydrological Plan, under which 272 new dams and aqueducts were intended to transport water from northern rivers to the arid south of the country to support a water-devouring horticultural export industry. In India, since the incumbent BJP government was similarly swept from power,

The grand hydraulic folly

In October 2002, at the height of a drought which was pitting states against each other over river allocations, the Indian Supreme Court issued a directive to the national government. A massive set of schemes long put on the backburner for environmental and technical reasons, should be resuscitated. Within ten years, India's major rivers should be interlinked.

The 'garland of rivers' proposal, whereby water will be pumped over highlands from river basin to river basin, flow through 9,600 kilometers of canals and be captured behind 32 major dams, is the ultimate expression of faith in concrete to solve the country's water shortage crises. The intention is to redesign the flows of 37 major river systems and transport water from the northern rivers to the south.

The Government dutifully set up a Task Force and went to work, but the subsequent furor has been immense. Environmentalists are appalled at this hydraulic folly on a giant scale. Fertile lands will become deserts, ecosystems will be destroyed, hundreds of thousands of villagers will be displaced, and tens of thousands of hectares of forests flooded. Since over 70 per cent of Indian river water is polluted, there would also be a spread of contaminants, damaging aquatic and human life. The costs – at $220 billion, twice the country's current annual national revenues – would cripple the economy with debt.

Indian states are already at loggerheads over different parts of the proposal. Punjab, supposedly a 'donor' of water, has already stated that it has none to spare, and Kerala is equally opposed to any transfer of its waters to other states. Bangladesh has protested internationally that diversions in upper reaches of the Ganges and Brahmaputra would threaten the livelihoods of millions of its citizens and irrevocably damage its coastal economy.

Since the elections in 2004 led to a change in government, there is hope that the new administration will quietly let the project drop. But since the original instruction came from the Supreme Court and the current Indian President APJ Abdul Kalam is behind it, the 'garland of rivers' may not vanish so easily. Some years of good monsoon would help kick it into the long grass, while the political and environmental wrangling over every feasibility study and sub-proposal may also help to pave the way for more democratic and practicable alternatives.

Information from many sources accessible at www.irn.org

activists have been pressing for the demise of the 'garland of rivers' project, a grandiose and controversial national plan for the interlinking of major river basins throughout the subcontinent.

Yet another context for water activism is ecosystem protection. The growing frequency of pollution scares; of bottled water and 'owning the rain' scandals; the rapid rate of groundwater depletion; the salinization of irrigated soil; the decline of wild fish and crustaceans; and the increased volatility of the weather due to climate change, have created fertile arenas for social and political protest. These skirmishes around water have helped raise its profile and improved public awareness of the various issues and their interconnectedness, and the interconnectedness of water issues to others – national security, employment, power generation, industrial growth. Some of the most effective environmental campaigns weave many of these strands together. One such example is the campaign for rainwater harvesting in India. This has managed to have rain collection and storage above and below ground accepted as part of water policy in parts of the country. In Tamil Nadu, for example, it is now mandatory for all buildings to have rainwater harvesting devices. The adherents of grandiose plans are still a major force. But perhaps not as immovable as they were.

Internationally, the campaign for basic drinking water and sanitation services 'for all' has also had some success, adding to the sense of heightened concern about survival and morbidity among the world's poorer inhabitants, especially in Africa. New commitments were wrung from governments and donors at the 2002 Earth Summit.

Despite activist gains, there is an unavoidable sense of unease. However often it is repeated that 'business as usual' cannot go on and that a new world water order is required, among the international and state bureaucratic apparatus nothing very much seems to change. If it does, the change is often peripheral or rhetorical. In April 2004 a consortium of international NGOs issued a scathing report on the OECD industrialized countries' failure to take the world's water crisis seriously and the shortfalls on funding commitments of which they are guilty. The crisis they were referring to was the lack of safe drinking water and sanitation for millions of people, and the interaction between this and other aspects of poverty such as poor health and low standards of education.

Such 'callings to account' of the international community for reneging on promises are welcome. But to fix upon domestic service delivery as the crux of the crisis is to do many of its victims a disservice. The problems are far wider and deeper than financing for drinking and flushing. Indeed, unless the whole framework of water law and policy is overhauled, and redesigned to ensure better management of the resource, in many places the rapid extension of services through the existing apparatus is impracticable. Where financial investments are poorly applied, people may end up no better and sometimes worse off than they were before.

For many people with access to wells and natural streams, 'unsafe' though they may be, it is important that they continue to have that access and do not find themselves excluded from water resources by something – a dam, a mechanized borehole, a bottling industry, a new organizational entity – masquerading as development advance but actually producing what, for them, is the reverse. For them, the incipient crisis is less the failure to hook up new yard taps, hand-pumps and pour-flush toilets – though all those are problems too; but the way in which access to water and to all natural resources is being gradually restricted to fully paid-up members of the modernizing and commodified world.

In the face of the bitter ideological divides concerning water resources and their management, how has the international water establishment performed?

The international record

Ever since the international community got together in the early 1990s to consolidate their views on water in advance of the first Earth Summit in 1992, they have pinned their colors to 'integrated water resources management' (IWRM) as the way to address the global freshwater crisis. There is nothing intrinsically wrong with this approach; in fact there is everything theoretically right with the idea of sharing, conserving and managing the entire water resources available – locally, nationally and globally – within an integrated framework. But it represents an ideal, and realizing the ideal is very remote when it comes to implementation on the ground. Everything rests on establishing a framework for sound 'water governance'.

Many of the policy principles laid down for integrated water resources management contain nothing to excite opprobrium even from the most skeptical water warriors and NGOs. There is an expressed commitment to equity and sustainability as a basis of service delivery. The preference for the river basin or catchment as the basis of water administration would be applauded by the most democratically inclined water activist: this is exactly the sort of organizational structure developed by local water user associations and water parliaments. There is a stated commitment to consultation from all stakeholders, with decision-taking at the lowest appropriate administrative level; and the need for community and women's participation, social mobilization, accountability and openness, are all underlined.

There are also mitigating circumstances for some of the new policy principles which have aroused such ire; for example, the insistence that water is an economic good. This is not really as sinister as it has been made to sound. The typical farmer, villager or pastoralist would agree – even if their terminology was different. Of course the value of water in crop production, in domestic supply systems and in industrial processes has to be calculated. Otherwise how can conservation regimes be devised based on encouraging low-volume as opposed to high-volume water consumption – for example, in selecting crop species, toilet designs, irrigation types and industrial processes? Equally, there is a role for graduated water tariffs and pricing, even for water and pollution trades in fully marketized environments, as part of the means of regulating demand and supply. It is useful, in suitable contexts, to use price incentives to penalize low-value uses – swimming pools and golf courses – as opposed to high-value uses – drinking and pollution control. If the state and its local extensions fulfil their custodial and regulatory role effectively, there is no reason to exclude entrepreneurship entirely.

So what actually is the problem? For there is a major problem. And it is the same problem which besets so much that is discussed to death in international forums, concerning water or anything else.

Look on the ground

Integrated water resources management and sound 'water governance' cannot exist in the ether. They can only exist on the ground.

The context in which policy principles have to be applied – in every case a unique mix of economic, social, cultural, hydrogeological, political, administrative and other environmental factors – determines which policies are suitable and whether they can be made to work. 'Good governance' cannot be invented as if it were a module or imported from outside; it needs its own roots and organic growth to flourish. Very little governance in poor societies is good or effective; it is usually under-resourced, inefficient, undemocratic and corrupt. There is no mystery about this, although the degree to which it is ignored implies that there is. It is simply a corollary of a country or areas within it being seriously 'underdeveloped'.

The water policy principles so painstakingly

thrashed out in Agendas and Guidelines may be excellent on paper. But practice on the ground falls desperately short. In many cases, even where key officials sign up to them in all sincerity, political support for the existing fragmented and irrational water regime is so entrenched that any expectation of realizing something radically different in the short or medium term is fanciful.

Because it is so difficult to make the principles work as they should – to get government to reform its entire water approach and exercise its custodial and regulatory role effectively – many donors renege even more heavily on the policy part of the agenda than on the financial side. Either they let governments off the hook, or they hand over responsibility to private sector bodies (companies and NGOs) in whom they have more confidence and can control by means of market forces and business arrangements. Accountability to project beneficiaries or participants is rarely given serious consideration except by NGOs. Many large-scale projects are not subjected to adequate scrutiny nor do donors use the leverage they possess to help effect the reforms which would make their international policy commitments meaningful.

Rarely do they examine closely and honestly on the ground the outcome of their partnership choices: the failures to anticipate environmental and social damage; the exaggerated estimates of cost-benefits by contractors whose profit motive is far stronger than their desire to provide a good service to customers with no means of redress; the lack of guarantees of rights of access to traditional herders, forest dwellers, fishing people and others whose interests and rights in the resource get overlooked. Above all, they do not take seriously the absence of social mobilization, democratic participation and appropriate technical choice which would have helped promote a process of social and economic change and given their projects some hope of real success.

Co-operating with centralized systems

What they do in practice therefore co-operates with the centralizing tendency whereby, instead of communities taking back control over their water affairs, residual power over their resource base tends to flow away. The defense of poor people's interests is left to NGOs, despite all the international rhetoric about 'goals' and financing targets. We never hear from government, industry, development bank or trade organization a resounding declaration in favor of rainwater harvesting, non-energized irrigation or 'dry' sanitation; nor do we see them open up a dialogue with existing providers with a view to building on locally-generated approaches.

No wonder the end-product of their engagement implies that the international water policy establishment has been co-opted by the water industry as a surrogate proponent of all the watery components of the neoliberal agenda which commoditize, deprive, marginalize and pollute: the very opposite of what IWRM is supposed to do. It may be unintentional, but the international bodies involved occupy a default position on the side of the ideological divide which favors centralized against local control of the resource base, national growth against sustainable livelihoods, technocratic mega-construction against small-scale eco-friendly initiatives, and privatization against public ownership of services.

The more distant the exercise of power over any common resource or service, the less control over it is wielded by people at the local level. This may not matter as much in a fully modernized market economy, but it matters a great deal in places where a large proportion of the population still depends heavily on the natural resource base for their livelihood, and where it is unrealistic to assume that a significant proportion of them can be absorbed into another economy which would improve life for them in the foreseeable future. Redistributional, pricing and regulatory systems which ought to redress the balance towards them if the state was properly fulfilling its responsibilities are more often dominated by bureaucratic, business or political interests whose purpose in holding power is to deploy these instruments to suit themselves.

Unless local leaders – official or NGO – manage to succeed in redressing the process in the favor of those faced with exclusion from modern economic advance, they stand no chance of enjoying equal rights of access to water resources, education, health services, markets, transport, electricity or anything else. Too often, the official policy environment is not on their side, even though there are a handful of heartening local exceptions.

Water is everybody's business

Many of the world's leading water warriors have set out the key components of a new world water order. If they have one thing in common, it is the notion that water as a vital natural resource should remain in common ownership, and that both the water and the pipes that conduct it about should be controlled by local democratic power, instead of by distant corporations with bureaucracies and businesspeople in their pockets. Apart from Vandana Shiva's advocacy of 'water democracies', another attempt to articulate a new world water order is Riccardo Petrella's 'world water contract' based on the founding principle that water is a vital common global heritage. The CSE in New Delhi echoes the idea of water's re-democratization in their slogan 'Making water everybody's business', and by their recommendation that people regain community control over their water resources by 'capturing the rain'.

Water issues are so diverse that they are difficult to define under an all-embracing view of a new world water order. We have already discovered that there can be no crude reduction to simple scenarios of free water services versus fully-charged, water for health versus water for food, farmer's rights versus environmental protection, water conflict versus water peace. But for many already engaged in the clash of civilizations between people's right to life and market optimization of every component of our natural wealth, water stands today for the entire global commons under threat.

First, historically, came the enclosure of land, the creeping extension – often by conquest or subterfuge – of title and ownership over all soil, earth, rocks, minerals, plant and animal life; then came the colonization of lakes, rivers and streams; now it's the turn of whatever so far survived in its natural state to be co-opted into the organized and marketized global maw by the intrusion of surveyors, proprietors, global investors and the rest. Where will the process end, ask those shocked by the possibility of farming people's, fishing people's and indigenous people's market-based exclusion from an essential life force. Does air come next?

What chance of water peace?

Thus the main issue surrounding water today is not its unfair distribution between those places which are naturally well-endowed with rain and those that are not. Instead it is the system of ownership and management of water which is supposed to compensate for natural disadvantage, but tends rather either to reinforce it or to introduce a whole new demarcation of water-rich and water-poor.

Water daily provides ever more opportunities of clash between dispossessed rural humanity and the onward march of market forces. So far, few of us have noticed because most of the clashes are far away and largely inaudible. We hear of starvation deaths, of deaths from flood, drought, and environmental catastrophe, of pollution sagas and epidemics of disease. We hear of people driven into poverty and marginalization by loss of their land, loss of their crops, loss of their jobs or their livelihood, and lack of access to services such as health care and education. But we don't recognize how central a role water plays in all these tragedies. It is time that we did.

Future water peace will depend less on international diplomacy between water-competing nations however important that is, than on whether water's status as a commons over which all humanity has rights can be upheld. That in turn depends on whether states and water administrators can be made to assume their proper custodial role towards the resource on which all living systems depend.

The signs are not altogether promising. But things can and do change. Perhaps water will manage to redeem the sacred status it has enjoyed since life appeared on earth, and magically divest itself of the commodity-driven character which is, after all, only a recent incarnation. If so, those around the world who have led the fight for water rights will deserve unstinting applause.

Contacts

International

Centre for Science and Environment
41, Tughlakabad Institutional Area,
New Delhi 110062, India
Tel: +91 11 29955124
Fax: +91 11 29955879
Email: cse@cseindia.org
Websites: www.cseindia.org
www.rainwaterharvesting.org

Freshwater Action Network
Website: www.freshwateraction.net

International Rivers Network
1847 Berkeley Way
Berkeley, CA 94703, US
Tel: +1 510 848 1155
Fax: +1 510 848 1008
Email: info@irn.org
Website: www.irn.org

One World International
Website: www.oneworld.net

Stockholm Environment Institute
Lilla Nygatan, 1
Box 2142, S-103 14 Stockholm, Sweden
Tel: +46 8 412 1400
Fax: +46 8 723 0348
Email: postmaster@sei.se
Website: www.sei.se

United Nations World Water Assessment Programme
UNESCO/Division of Water Sciences
1 rue Miollis
75015 Paris, France
Fax: + 33 1 45 685811
Email: wwap@unesco.org
Website: www.unesco.org/water/wwap

Water Supply and Sanitation Collaborative Council
International Environment House, Chemin des Anémones 9, 1219 Châtelaine, Geneva, Switzerland
Website: www.wsscc.org/

Aotearoa/New Zealand
Oxfam Water for Survival Programme
Website:
www.oxfam.org.nz/Water/index.htm

Australia
OzGREEN
PO Box 1378
Dee Why, NSW 2099

Tel: +61 2 9984 8917
Fax: +61 2 9981 4956
Website: www.ozgreen.org.au

Water Matters Campaign
PO Box 164, Blackburn VIC 3130
Website: www.watermattersaustralia.org

Canada
The Council of Canadians
501-151 Slater Street
Ottawa, Ontario, K1P 5H3
Tel: +1 800 387 7177
Website: www.canadians.org

WaterCan
Website: www.watercan.com

UK
WaterAid
Prince Consort House
27-29 Albert Embankment, London SE1 7UB
Tel: +44 20 7793 4500
Fax: +44 20 7793 4545
Website: www.wateraid.org

US
Pacific Institute
654 13th Street
Preservation Park, Oakland, CA 94612
Tel: +1 510 251 1600
Websites: www.pacinst.org
www.worldwater.org

Bibliography
see notes at the end of each chapter

Internet Resources
Friends of River Narmada:
www.narmada.org
International Institute for Environment and Development: www.iied.org
Operation Water Rights at the Polaris Institute: www.polarisinstitute.org
Research Foundation for Science, Technology and Ecology, India:
www.vshiva.net
Right to water:
www.righttowater.org.uk
United Nations Environment Programme:
www.unep.org/themes/freshwater
World Commission on Dams:
www.dams.org
World Health Organization water section:
www.who.int/health_topics/water/en/
Worldwatch Institute:
www.worldwatch.org

Index

143

Index